NANOTECHNOLOGY SCIENCE AND TECHNOLOGY

NANOSCIENCE, NANOMANUFACTURING AND NANOELECTRONICS

NANOTECHNOLOGY SCIENCE AND TECHNOLOGY

Additional books in this series can be found on Nova's website
under the Series tab.

Additional E-books in this series can be found on Nova's website
under the E-book tab.

NANOTECHNOLOGY SCIENCE AND TECHNOLOGY

NANOSCIENCE, NANOMANUFACTURING AND NANOELECTRONICS

JOSEPH G. DEVIN
EDITOR

Nova Science Publishers, Inc.
New York

For permission to use material from this book please contact us:
Telephone 631-231-7269; Fax 631-231-8175
Web Site: http://www.novapublishers.com

LIBRARY OF CONGRESS CATALOGING-IN-PUBLICATION DATA

Nanoscience, nanomanufacturing, and nanoelectronics / editor, Joseph G. Devin.
p. cm.
Includes bibliographical references and index.
ISBN 978-1-61761-816-1 (hardcover)
1. Nanotechnology. I. Devin, Joseph G.
T174.7.N3584 2010
620'.5--dc22

2010036418

Published by Nova Science Publishers, Inc. † New York

CONTENTS

PREFACE

Nanotechnology is the study of the controlling of matter on an atomic and molecular scale and is also very diverse, ranging from extensions of conventional device physics to completely new approaches based upon molecular self-assembly, from developing new materials with dimensions on the nanoscale to investigating whether we can directly control matter on the atomic scale. This book discusses nanostructures and their properties which are critical to understanding and developing innovations in biological systems, therapeutic agents, and medicine and health.

Chapter 1- Nanostructures and their properties are critical to understand and develop innovations in biological systems, therapeutic agents, and medicine and health. However it has only been in the last five years that “nanomedicine” as a field has been created and has rapidly accelerated. This paper explores the reasons behind that fact, examines the science and engineering issues that remain to be addressed if one is to more rapidly translate nanoscience discovery into nano-enabled medical technology, and suggests Federal agency actions that could accelerate that eventuality.

The U.S. National Nanotechnology Initiative was instituted in 2001 to accelerate and exploit progress in the science and engineering of nanostructures. As evident in Figure 1, exponential growth in literature addressing the nanoscale began about 1990. Interest in the nanoscale has been driven by the commercial availability of nanoscale manipulation and characterization tools, the expectation of new physical, chemical, and biological properties of nanostructures, the expectation that nanostructures will provide new building blocks for innovative new materials with novel properties, the miniaturization into the nanoscale by the semiconductor industry, and the recognition that the molecular machinery in a biological cell functions at the nanoscale. Historically, aspects of chemistry and biology - such as colloids, protein engineering and molecular virology - have involved nanostructures, but on a largely empirical basis. Finally, there is an expectation that a better understanding of the 1-100nm materials size scale (the nanoscale) will lead to a seamless integration of theory and models across the size scales that encompass atomic-molecular-nanostructure-microstructure behavior and thereby enable the *a priori* prediction and design of a material’s properties.

Chapter 2- An increasingly competitive global market is driving industry to identify more effective ways to meet customer needs by adding value and reducing costs. Nanotechnology offers the promise of dramatically increasing competitiveness by removing long-standing performance limitations. In addition, nanotechnology is expected to revolutionize manufacturing processes and products across many high-value industry sectors, such as medical technology, electronics, aerospace, automotive, chemical, food and dietary supplements, forest products, and pharmaceuticals. Global investment in nanotechnology rose to nearly $13.5 billion in 2007, with a global market predicted by Lux Research to grow from $147 billion in 2007 (see Exhibit 1-a) to over several trillion dollars in the next decade. But significant technical barriers inhibit the discovery, high-volume manufacture and widespread commercialization of nanotechnology-enabled products needed to fuel this growth. The United States Government has invested over $10 billion over the last ten years and is a long-standing leader in nanotechnology funding. Other nations are now rapidly increasing their investments and eroding this lead.

Chapter 3- It has been suggested that by 2025 at least half of the companies comprising today's Dow Jones Index will be replaced by nanotechnology companies. In a recent report, Lux Research forecast the worldwide market for nanotech products and services to reach $2.6 trillion by 2014. If this happens, it will represent approximately 15% of global manufacturing output.[1]

Nanoelectronics is defined as advanced electronic and optical systems that depend upon innovations in nanotechnology. The field of nanoelectronics embraces a range of technologies, all ultimately for the purpose of creating and improving the functionality, cost, and performance of tomorrow's electronic systems. This chapter examines the historical role the Advanced Technology Program (ATP) has played in fueling innovation and American competitiveness in areas of nanotechnology that will lead to totally new-to-the-world electronic and optical products.

In: Nanoscience, Nanomanufacturing and Nanoelectronics ISBN: 978-1-61761-816-1
Editor: Joseph G. Devin

Chapter 1

RE-ENGINEERING BASIC AND CLINICAL RESEARCH TO CATALYZE TRANSLATIONAL NANOSCIENCE

James S. Murday[1], Richard W. Siegel[2], Judith Stein[3], J. Fraser Wright[4],

[1]University of Southern California
[2]Rensselaer Polytechnic Institute;
[3]General Electric Global Research;
[4]Children's Hospital of Philadelphia and University of Pennsylvania

I. INTRODUCTION

Nanostructures and their properties are critical to understand and develop innovations in biological systems, therapeutic agents, and medicine and health. However it has only been in the last five years that "nanomedicine" as a field has been created and has rapidly accelerated. This paper explores the reasons behind that fact, examines the science and engineering issues that remain to be addressed if one is to more rapidly translate nanoscience discovery into nano-enabled medical technology, and suggests Federal agency actions that could accelerate that eventuality.

The U.S. National Nanotechnology Initiative was instituted in 2001 to accelerate and exploit progress in the science and engineering of nanostructures. As evident in Figure 1, exponential growth in literature addressing the nanoscale began about 1990. Interest in the nanoscale has been driven by the commercial availability of nanoscale manipulation and characterization tools, the expectation of new physical, chemical, and biological properties of nanostructures, the expectation that nanostructures will provide new building blocks for innovative new materials with novel properties, the miniaturization into the nanoscale by the semiconductor industry, and the recognition that the molecular machinery in a biological cell functions at the nanoscale. Historically, aspects of chemistry and biology - such as colloids, protein engineering and molecular virology - have involved nanostructures, but on a largely empirical basis. Finally, there is an expectation that a better understanding of the 1-100nm materials size scale (the nanoscale) will lead to a seamless integration of theory and models across the size scales that encompass atomic-molecular-nanostructure-microstructure behavior and thereby enable the *a priori* prediction and design of a material's properties.

The nanoscale literature in the 1990s is dominated by investigations of "hard" materials – ceramics, metals, semiconductors - only in small part due to the keen interest in nanoelectronics devices. Many of the new nanoscale analytical tools depend on proximity between a tip and the sample under investigation. This requirement was not overly onerous for relatively stiff materials, the primary focus of nanotechnology in the 90s. In contrast, soft materials, those of predominant interest in the biology and medical communities, are more readily deformed by a proximal probe and are thereby more difficult to analyze quantitatively. It wasn't until roughly 2000 that improvements in commercially available instrumentation made the analysis of soft material more viable. Not coincidently, the literature reporting nanostructures in biology, medicine and health began to increase more rapidly at that point (see Figure 1).

The increase in research activity has to be complemented by programs in the various funding entities. For example, in the U.S. there are eleven agencies with nanoscale research and development (R&D) programs reported by the U.S. National Nanotechnology Initiative (NNI, see Table 1) [1, 2]. The reported funding for nanoscale science and technology has grown by nearly a factor of 4 since the initiation of the NNI in 2001.

At the beginning of the NNI, the National Institutes of Health (NIH) investment at the nanoscale was modest. The NIH hosted two workshops - Nanoscience and Nanotechnology: Shaping Biomedical Research, June 2000 [3], and Nanobiotechnology, October 2003 [4] - to better understand the potential impact of nanostructures on medicine/health and the knowledge deficiencies inhibiting progress. These workshops, along with promising results from research [5], led to major increases in the NIH investment in nanoscale research. The nanoscale investment by NIH has more than doubled since 2005 (sextupled since 2001 – see Table 1).

As part of the NIH investment to exploit the nanoscale, Nanomedicine was incorporated into the NIH Roadmap for Medical Research in 2004 [6]. Understanding nanoscale properties permits engineers to build new materials structures and use these materials in new ways. The same holds true for the biological structures inside living cells of the body. To meet the challenges, and complement its Institute-based programs, the NIH established a national network of eight Nanomedicine Development Centers. These collaborative centers are staffed by multidisciplinary research teams including biologists, physicians, mathematicians, engineers and computer scientists. In the initial phase of the program (FY2005-FY2010), research has been primarily directed toward gathering extensive information about the chemical and physical properties of nanoscale biological structures.

The European Science Foundation launched a Scientific Forward Look on Nanomedicine in 2004, which involved a series of five workshops and a Consensus Conference (Nov 2004). This was followed in Nov 2006 by a report with a strategic research agenda for nanomedicine [7]. Founded in 2007, the European Society of Nanomedicine [8] shares office space with the European Foundation for Clinical Nanomedicine (CLINAM foundation) [9]. The first European Conference for Clinical Nanomedicine was organized by CLINAM in May 2008 and had sessions on unsolved problems waiting for nanomedical solutions, nanotechnologies at hand for solving medical problems, clinical trials in nanomedicine, and building bridges between clinicians and nanoscientists. A proceedings from the NATO Advanced Research Workshop on Nanomaterials for Application in Medicine and Health has been published [10].

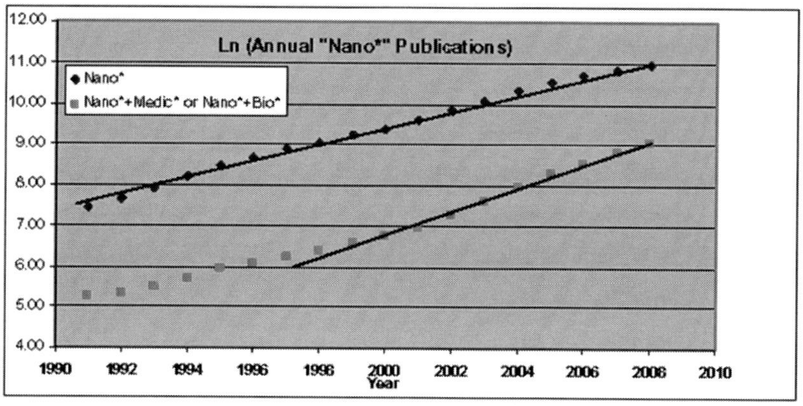

Figure 1. Publication counts derived from the Thompson ISI Web of Science database on 3/15/2009 using the indicated keywords. The vertical axis is the natural logarithm of the number of publications. There is a clear change in slope for the publications associated with biology and medicine around the year 2000.

Table 1. U.S. Federal Agency R&D Funding in the NNI

	FY2001 requested*	FY2001 actual*	FY2005 requested*	FY2005 actual*	FY2009 requested*
NSF	217	150	305	335	431
DOD	110	125	180	352	397
DOE	94	88	211	208	311
HHS (NIH, NIOSH)	36	40	89	168	232
DOC (NIST)	18	33	53	79	110
NASA	20	22	35	45	6
EPA		5	5	7	5
USDA (CSREES, FS)			5	2	8
DOJ			2	2	2
DHS (TSA)			1	1	1
DOT (FHWA)					1
Total		~460		~1200	~1500

* The Presidential Budget submission presents requested funding; actual funding is reported by the agencies after the end of the fiscal year. The differences reflect Congressional appropriation decisions, including Congressional adds, and agency funding decisions taken during the fiscal year.

The growing attention to nanostructures in medicine/health is also reflected in the professional science and engineering communities. A Handbook of Nanomedicine is being implemented [11]. The American Academy of Nanomedicine, founded in 2005, launched a journal - Nanomedicine: Nanotechnology, Biology and Medicine (ISSN1549-9634). The Institute of Nanotechnology, founded in the UK in 1994, began its nanomednet [12] in 2007. Additional journals, the International Journal of Nanomedicine (ISSN 1176-9114) and Nanomedicine (ISSN 1743-5889) were both launched in 2006. While not strictly "nano," another relevant journal is the Royal Society of Chemistry's Lab on a Chip [13] initiated in

2001. The American Physical Society PACs classification scheme has been modified to include "nanotechnology design" (87.85.Qr) and "nanotechnology application" (87.85.Rs) under Biomedical Engineering (87.85). Recurrent professional forums addressing aspects of nanomedicine are: The International Nanomedicine and Drug Delivery Symposium (NanoDDS – began in 2003) [14], the Annual Meeting of the American Academy of Nanomedicine (began in 2005) [15], the Society for Biomaterials meeting [16], the AVS International Symposium and Exhibition [17], the European Science Foundation conference Nanomedicine 2008 [18], and the International Conference on Biomedical Applications of Nanotechnology [19].

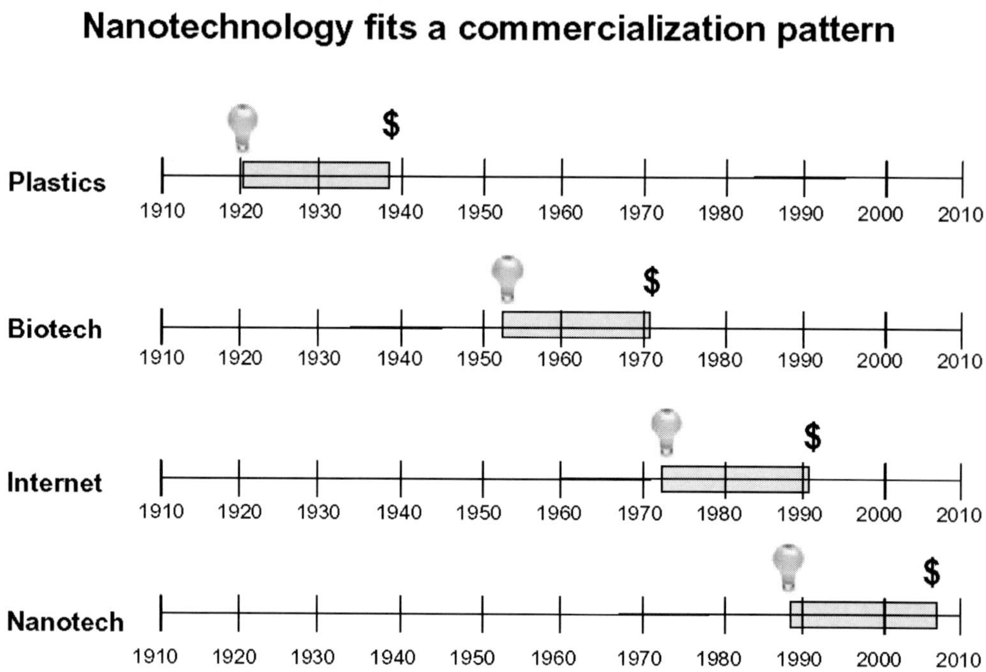

Figure 2. Time required for maturation of science discoveries into commercial products. Courtesy Lux Research Inc. • One Liberty Square, Suite 210, Boston MA 02109

The paucity of knowledge for nanostructure impact on environment, safety, and health (ESH), coupled with the growing prevalence and diversity of, and especially the novel engineered properties in, nano-enabled technologies continues to raise ESH concerns [20]. Knowledge of nanostructure ESH risks and their amelioration will be symbiotic with health and medicine applications since understanding how to avoid health problems can potentially be used to guide therapy and *vice versa.* Workshop and task force reports on ESH issues include: Nanotechnology: A Report of the US Food and Drug Administration [21], and Environmental, Health, and Safety Research Needs for Engineered Nanoscale Materials [22], a UK report on Nanomaterials risk [23], an International Council on Nanotechnology (ICON) report [24], and a European Commission paper [25].

Conventional wisdom, buttressed by observation, posits a 20 year gestation period between science discovery and its exploitation in the market. Figure 2 illustrates this point for several major 20th century technologies, and suggests imminent emergence of nano-enabled technologies. In practice, nanostructures have been utilized in selected technologies for some time, including carbon black in tires, colorants in stained glass windows, colloidal silver as a disinfectant, colloids and colorants in cosmetics, and many catalysts. Small particles have also been in use for biomedical research and *in-vitro* diagnostic protocols during the last fifty years [26, 27]. Most of these applications are based on technology derived empirically from macroscopic observations.

There is a database with a listing of products in the market that are (or claim to be) nano-enabled [28]; there are also specific listings of nano-enabled products in medicine and health [29, 30]. As the capability to make, measure, and manipulate individual nanostructures continues to progress, one can expect more extensive entry of nano-enabled and nano-enhanced products and technologies into the marketplace in the coming years.

While twenty years might be the generic interval for transition of science discovery to innovative technology to translation of these technologies into medical applications, a thoughtfully crafted investment strategy might shorten this time frame. To improve human health, scientific discoveries must be translated into practical applications. Such discoveries typically begin at "the bench" with basic research — in which scientists study disease at a molecular or cellular level — then progress to the clinical level, or the patient's "bedside". Scientists/engineers are increasingly aware that this bench-to-bedside approach to translational research is really a two-way street [31]. Basic scientists provide clinicians with new tools for use with patients and for assessment of their impact; clinical researchers make novel observations about the nature and progression of disease that often stimulate basic investigations. Translational research has proven to be a powerful process that drives the clinical research engine [32]. However, a stronger research infrastructure could strengthen and accelerate this critical part of the clinical research enterprise. Through discussions with deans of academic health centers, recommendations from the Institute of Medicine, and meetings with the research community, the NIH recognized that a broad re-engineering effort is needed to create greater opportunity to catalyze the development of a new discipline of clinical and translational science. An outcome was the launch of the Clinical and Translational Science Awards (CTSA) Consortium in October 2006.

Accelerating discovery into technological device is certainly a goal for the DOD hierarchy of research and development funds (evolving from basic (6.1), applied (6.2), advanced technology development (6.3), advanced component development and prototypes (6.4), to system development and demonstration (6.5)) and the NASA research and development funds evolving from technology readiness levels 1 through 6. One might consider the NIH equivalent as R01 programs to explore new concepts (basic), the R21 programs to demonstrate feasibility (applied research), the R33 programs to further develop the idea toward a technological/methodological goal, and the Clinical and Translational Science Awards (CTSA) providing translation into the clinic (which might be considered the equivalent of initial field testing by DOD or NASA).

Many agencies, NIH included, also utilize SBIR/STTR and public-private partnerships as a bridge into commercialization. In contrast to many DOD/NASA technologies, because there are large public markets for medical/health technologies, commercialization does not have to rely as predominantly on additional Federal funding.

The April 2008 report on the NNI by the Presidents Council of Advisors on Science and Technology (PCAST) emphasizes the central role the NNI must play in overcoming the barriers to nanotechnology development and commercialization [33]. The 2009 U.S. NNI reauthorization draft bill (H.R.554, 111th Congress) also places greater emphasis on applications. As a key supporter of nano-bio basic research, NSF sponsored a workshop on Re-Engineering Basic and Clinical Research to Catalyze Translational Nanoscience on 16-19 March 2008, hosted by the University of Southern California, with a goal to provide insights toward a Federal government nanomedicine investment strategy. The workshop participants, listed in Table 2, were carefully chosen to reflect both individuals working in nanoscience and nanoengineering and those working in clinical settings. The NIH - Institute of Biomedical Imaging and Bioengineering (NIBIB) sponsored a conference (20-21 March 2008) to present the preliminary findings.

Table 2. Workshop Participants

Name	Institution
Mark Braganza, MD	Texas Pacific Group Growth
Tom Buchanan, MD	University of Southern California
Wah Chiu, PhD	Baylor College of Medicine
Vicki Colvin, PhD	Rice University
Richard Cote, MD	University of Southern California
William Galey, PhD	Howard Hughes Medical Institute
Martha Gray, PhD	Harvard University
James Heath, PhD	California Institute of Technology
James Hone, PhD	Columbia University
Mark Humayun, MD, PhD	University of Southern California
Anupam Madhukar, PhD	University of Southern California
Ellis Meng, PhD	University of Southern California
Michael Roukes, PhD	California Institute of Technology
Jeffrey Schloss, PhD	NHGRI, NIH
Richard Siegel, PhD	Rensselaer Polytechnic Institute
Judith Stein, PhD	General Electric Corporation
Edwin Stone, MD, PhD	University of Iowa
Samuel Stupp, PhD	Northwestern University
Sally Tinkle, PhD	NIEHS, NIH
Fraser Wright, PhD	Children's Hospital of Philadelphia and Univ Penn.
Steven Moldin, PhD	University of Southern California
James Murday, PhD	University of Southern California

Based on information derived from the workshop and other resources, the succeeding sections of this chapter will examine opportunities to accelerate the development and optimization of nanostructures for impact on medicine and health. The next section briefly explores economic and societal drivers for nanomedicine initiatives. The third section will examine the science, engineering, and medical research needs. The fourth section succinctly examines the U.S. Federal investment directly germane to medicine and health, with brief mention of the European Union (EU) effort. The final section presents recommendations to accelerate the translation from laboratory discovery into clinical practice.

II. HEALTHCARE NEEDS WITH PROMISING NANO-ENABLED TECHNOLOGY IMPACT

An investment strategy to accelerate nanoscience into nano-enabled technology for medicine and health should reflect both health need (technology pull) as well as science push. Within this context, science and engineering of nanoscale structures is expected to make major contributions across the entire medicine and health spectrum ranging from mortality rate, morbidity an illness imposes on a patient, disease prevalence, and general societal burden [29, 34-36]. The following examples illustrate potential economic and therapeutic impacts, even if nano-enabled technologies only contribute partial solutions:

- The direct medical cost for cancer in the U.S. for 2007 was about $90B [37]. The National Cancer Institute has recognized the importance of nanostructures in the diagnosis and treatment of cancer in its Alliance for Nanotechnology in Cancer (http://nano.cancer.gov/). Nanotechnology approaches [38] are progressing rapidly in: early diagnosis [39-41], nano-enabled contrast agents for in-vivo imaging [42-47], nano-reformulations of chemotherapy agents for lesser quantities of drug, targeted delivery for smaller side-effects [48-52], and new treatments such as nanoparticle-mediated tumor ablation [5, 53].
- The direct medical cost for diabetes in the U.S. for 2007 was about $116B [54]. Nanotechnology approaches to *in-vivo* monitors of glucose levels [55, 56] and production of insulin [57] are being explored.
- The annual medical care cost for spinal cord injury in the U.S. is about $1.5B; the full costs are estimated as about $10B/yr [58, 59]. There are promising nano-enabled approaches to the regeneration of spinal neurons, a capability once thought impossible [60, 61].
- In the US, in order to remain physically active, approximately 200,000 people receive hip implants and 300,000 people receive knee implants [62]. The average life-time of current orthopedic implants is only 10 – 15 years; revision surgeries and their recoveries are not as successful as the first operation. The cost of an implant varies but is roughly $20,000. Nano-enabled innovations in bone cement and composite structures are opening new possibilities for improvements in implants [63-66].

The 2006 "Nanomedicine: Nanotechnology for Health" publication of the European Technology Platform – Strategic Research Agenda for Nanomedicine presents additional examples of expected nanomedicine impact [67].

The workshop participants were polled for their opinion of pressing clinical needs amenable to nano-enabled technology and identified the following as illustrating the wealth of opportunities:

- Intelligent nanobiomaterials for cell therapy to improve heart function
- Safe and affordable therapeutic strategies to regenerate neural tissues
- Kidney – hollow fiber membranes
- Detoxification implants- correction of metabolic disorders

- Cochlear and retinal implants
- New power source technology for implants
- Repair articular cartilage and regain homeostasis with the joint
- Skin regeneration
- Anti-microbials
- Drug delivery with
    - Targeted pharmacotherapy – tissue/organ
    - Therapeutic DNA transfer vectors
    - Nanoparticle (NP) to carry a therapeutic payload across the blood brain barrier
    - Transfection devices for therapeutic uses.
    - Controlled release (especially long term, continuous and programmed)
    - Transient application – sonoporation and electroporation

III. Nanoscale Science and Engineering Research Needed to Enable More Effective Technology

The considerable investment in nanoscience across the world has been leading to many new discoveries. In the second five years of the U.S. National Nanotechnology Initiative there is growing effort to identify potential applications for those discoveries and to accelerate their transition into innovative technology solutions to societal problems. Medicine and health provide fertile ground for this goal. For convenience, this section is organized about the headings of Diagnostics, Drugs, and Therapy, Implants and Regeneration, Systems Biology, and Medical Instrumentation and Devices. The amount of published work is growing rapidly; this paper's intent is not to be exhaustive, but rather illustrative of these topics.

IIIa. *In vivo* and *In vitro* Diagnostics

IIIa1. In-vivo Imaging – Contrast Agents

A number of non-invasive medical imaging approaches - such as computed tomography (CT), magnetic resonance (MR), positron emission tomography (PET), single photon emission CT(SPECT), ultrasound (US) and optical imaging (OI) - are currently being used. The emergence of nanosized contrast agents for these tools has been the subject of several recent reviews [45-47, 68, 69] and is anticipated to lead to advancements in imaging for understanding biological processes at the molecular level. Examples of these nanoparticles are biocompatible polymer-based nanogels/nanospheres/nanoemulsions, carbon nanotubes, dendimers, gold nanoparticles, liposomes, micro-bubbles, semiconductor quantum dots, silica nanoparticles with enclosed fluorescers, and superparamagnetic iron oxide particles. Generic goals for these contrast agents include:

- Signal-to-noise (S/N) enhancement to allow high sensitivity and resolution levels.
- Selective binding to target cells to provide a localized, specific enhancement.
- Long circulating half-life (hours) to expand the imaging time window.

- Acceptable toxicity profile.
- Ease of production and clinical use in order to be economically and commercially sustainable.

Nanostructured imaging contrast agents are overcoming many limitations of conventional contrast agents such as poor photostability, low quantum yield, and insufficient *in vitro* and *in vivo* stability. They are small enough to be taken up by single cells - via processes such as phagocytosis, pinocytosis or vector-mediated transport - as labels for *in vivo* imaging [42, 43]. Since the nanostructures are sufficiently small, one can envision linking them to provide multiple functions, opening possibilities for multimodal imaging, high payload of imaging reporter, activatable "on-off" systems, and chemical information. For instance, human oral cancer cells have been found to assemble and align gold nanorods conjugated to anti-epidermal growth receptor antibodies [70]. Molecules near the nanorods on the cancer cells gave a Raman spectrum (SERS) that was enhanced, sharp and polarized; those spectral features could be used for diagnostic signatures.

Research issues and opportunities include:

- Controlling nanomaterials themselves: a) purity, b) particle size and shape, and c) size distribution (mono-dispersity) of the particles. Their characterization requires specialized laser scattering techniques and microscopies to measure particle size and overall morphology, core composition/structure techniques, and surface analysis for composition/structure of any shells. The high-end instruments necessary for this characterization may not be readily available for the normal research laboratory and may require the use of user facilities or private analytical groups.
- Exploiting new contrast mechanisms - such as nanoparticle enabled surface-enhanced Raman [68, 71-73] - for the extraction of molecular spectral information.
- Ascertain the effects of composition, size, coating, surface charge density and the attached ligands on nanostructure pharmacokinetics and biodistribution.
- Explore mechanisms to ensure that particles are not rapidly accumulated in the spleen and liver.
- Delivery of nanoparticles to the cytoplasm of live cells.
- Incorporate multiple targeting ligands for enhanced selectivity. More than one epitope may be over expressed on a cell surface at a given time, so heteromultivalent probes for diagnosis and therapeutics will likely be important to selectivity.
- Utilizing nanoparticles as building blocks to obtain multimodal functionality – such as recognition, enhanced contrast, functional imaging, and therapeutic action.

IIIa2. In-vivo Diagnostics (non-imaging)

As microelectromechanical systems (MEMS) become more sophisticated and miniaturization continues into the nanoscale (MEMS evolving into nanoelectromechanical systems, NEMS), it will become possible to incorporate ever more sophisticated analytical capability onto and into the human body [74, 75]. There are already a growing number of miniaturized devices for transdermal sensing. Examples include the SCRAM system which is a high-tech bracelet that samples a person's sweat to monitor alcohol ingestion [76]; Echo's Symphony™ tCGM System which is a non-invasive (needle-free), wireless, transdermal continuous glucose monitoring (tCGM) system [77]; and Flexible Medical Systems, which is testing a MEMS chip with readable via RFID technology that could be applied to the skin via a band-aid to sense body fluid constituents [78]. As miniaturization continues, enabling greater sophistication per unit volume, the variety of measured analytes (and other properties such as temperature, tissue turgidity, etc.) will certainly increase. It is also within the realm of possibility that "spectrometers-on-a-chip" will be developed for insertion beneath the skin.

Research issues and opportunities include:

- Continued miniaturization for more sophisticated sensing of medically relevant parameters, including nanostructures for sensitive and selective transduction of biological events into electrical signals.
- Development of more effective wireless communications and power delivery suitable for the human body and human body exposures.
- Additional items as listed in Section IIIc, Implants.

IIIa3. In-vitro Miniaturized Diagnostics

Current approaches to medical diagnostics are usually high-cost, bench-top laboratory analyzers, or disposable kits that only test for a single analyte. The challenges are considerable when determining whether sophisticated chemical/physical laboratory instrumentation can be reduced in size, while retaining adequate capabilities. For instance there are 10^4-10^5 different proteins in blood with concentration ranges from 10^{-3} to 10^{-17}M. Miniaturized chip-based, array detection methods, known as microarrays, have been prevalent in almost all areas of health-related research for some time; continued miniaturization into nanoarrays will generate many orders of magnitude increase in multiplexed detection [79, 80]. Further, as noted in the section on contrast agents, nanoparticles are already incorporated into some diagnostics to provide greater selectivity, sensitivity, and practicality as compared with conventional systems.

Beyond the arrays, an alluring potential for micro/nano devices is to harness the concept of laboratory-on-a-chip for medical diagnostics [81-92]; analytical microchips are considered to be a fast growing technology [93]. The lab-on-a-chip concept (incorporating microfluidics) has several features that have attracted users in biology, chemistry, engineering and medicine. It requires only small volumes of samples and reagents, produces little waste, offers short reaction and analysis times, is relatively cheap, and has reduced dimensions compared with other analytical devices. Potential applications include: point of care measurements of saliva for periodontitis[41], heart disease [94] , hemacrit determination [95], insulin detection [96], and improving healthcare accessibility [97]. In addition to improving established diagnostics, new approaches - such as mechanical analysis to distinguish cancerous cells from normal ones even when they show similar shapes [98] - will be discovered and implemented. The lab-on-a-chip concept is particularly attractive as an approach to providing inexpensive, effective medical care especially in underserved populations [40, 99, 100].

Research issues and opportunities include:

- Arrays that can carry out larger numbers of experiments in parallel to assess reproducibility.
- Separation techniques to sort body fluid constituents.

- Optical and electromagnetic technologies incorporated in chip-based devices for manipulating samples and their analysis.
- The need for improved knowledge of pertinent biomarkers so that chip technologies can be more effective.
- More robust alternatives to antibody/antigen for selectivity determination.
- The need for validated and easy to operate microfluidic platforms which give the users the freedom to easily combine the basic modules for different fluidic operations in order to build application-specific microfluidic systems [101].

IIIb. Drugs, Delivery, and Therapy

IIIb1. New Approaches to Drug Development

There is need to analyze potential drug candidates in a more rapid and accurate manner, a need that provides an opportunity to develop new tools for that purpose. Ideally, to provide specific information, quantification of single cell pharmacokinetics/dynamics is desired, but requires the detection of minute quantities of proteins and other molecules [102]. One must identify and evaluate types of drug targets (including proteins, polysaccharides, lipids and nucleic acids) that can interact with small-molecule therapeutic agents. Polysaccharides, lipids and nucleic acids have been investigated less frequently than proteins, because of a lack of understanding of the involvement of these molecules in disease and a lack of small molecule therapeutic agents.

Traditional high throughput systems (HTS) perform by using multiple-well plates. Microfluidic technologies, and their nano-enabled enhancements, have great potential in high-throughput studies involving target selection, lead compound generation, identification, and dosage design.

In general, in-vivo imaging has been focused at the diagnostics level and has not been an inherent component of drug discovery. In the past decade a paradigm shift has occurred; imaging is now adding a new dimension to our understanding of basic biological and pharmacological mechanisms [103]. Many aspects of drug development can be facilitated using molecular imaging as an integrative tool to discover new "druggable" targets, identify novel drug candidates and validate their potency, sensitivity, specificity, PK, PD and toxicity, metabolism and adverse drug-drug interactions in living systems. Clinical, epidemiological and bioinformatical data suggest that population-wide genetic polymorphism may dictate the responsiveness to molecular therapy. Novel drugs are envisioned to be specifically tailored to selected patients.

While the greatest use of nanotechnology to date has largely been in passive carriers for drug delivery (see next section), nanoparticles themselves hold potential as therapeutic agents [104, 105]. To date, about ten years after the regulatory approval of liposomally encapsulated doxorubicin to treat various forms of cancer, "higher functionality" nanoparticles such as gene transfer vectors are in the investigational new drug stage of clinical research [106]. On the horizon is the use of nanoparticles themselves as drugs – as truly active "nanomedicines".

Research issues and opportunities include:

- Continued miniaturization of microtiter plates to expand the number of materials being evaluated.
- Dispensing nanoliter volumes of liquid into the wells.
- Microfluidic devices for drug screening that are sufficiently simple and highly versatile to enable their use in both academic and industrial pharmaceutical labs.
- Drug screening chips that incorporate living cells.
- Tailored nanoparticles as drugs

IIIb2. Innovative Drug Delivery

Nanotechnology advances are the cornerstone of a paradigm shift in targeting and safely delivering agents - thereby improving controlled drug release, improving patient safety and compliance, and reducing side effects [107-112]. Through the use of colloid chemistry – e.g., liposome and micelle encapsulation – nanoparticles have been used for drug delivery for decades. Examples of nanoscale delivery vehicles now under investigation include polymeric particles [113, 114], dendrimers [48, 115], nanoshells [116, 117], liposomes [51, 118], and magnetic nanoparticles [26, 119]. There are a number of new delivery platforms in clinical trial including: a dendrimer-derived microbiocide (i.e. VivaGel, Starpharma) for HIV or genital herpes in its final stage of Food and Drug Agency (FDA) approval[48], and a dendrimer-based targeted delivery of chemotherapeutic drugs and an apoptotic sensor in cancer cells [120]. Nanoparticles also show considerable promise for drug delivery to the retina and for powering prosthetic "artificial retinas" [121]. Various gene delivery systems based on nanoparticles have been developed and different polymers have been tested as gene delivery agents [122]. Going beyond simply carrying a drug, the development of "smart" nanoparticles is an exciting and promising area of investigation, [39, 49, 52, 123-125].

Delivering intravenous agents to their intended targets is no easy task. For intravenous infusion it is estimated that only approximately one of every 100,000 molecules of agent reaches its desired destination. The generic requirements of delivery systems are[49]:

- Biodegradability and biocompatibility.
- Stability under the *in vivo* conditions.
- In the bloodstream, it must be withdrawn from the circulation where the pathology is located, to reach elevated drug concentrations in the target cells.
- Allow sustained release of the drug, to achieve therapeutic levels at the site of pathology over long periods of time.
- Prevent the drug from manifesting its pharmacological and toxicological actions until present at the site, hence decreasing the incidence of side effects.
- Prevent premature degradation of the encapsulated drug and also immunological reactions.
- Minimize resistance due to low drug permeation levels in bacteria and phagocyte cells.

Nanoscale building blocks provide opportunity for multifunctional packaging small enough to navigate body vessels and membranes. A multifunction approach is needed to circumvent the body's natural defenses or biobarriers, which act as obstacles to foreign objects injected in the blood stream. One must avoid them being removed from circulation by monocytes and macrophages or accumulated in the organs of the reticulo-endothelial system (RES), especially the liver and spleen. For instance, for many cancers endothelial gaps in tumor vasculature are measured in hundred of nanometers rather than in tens of nanometers. In this case, nanocarriers in the appropriate size range could more selectively extravasate into a tumor and provide a passive means for selective delivery [126]. Other factors influencing the magnitude and pattern of tumoral distribution are [127]: *in-vivo* stability, particlerface charge, and intracellularsize, su uptake.

In most cases active selectivity will be desirable. Targeted drug delivery carriers are being functionalized with antibodies or antibody fragments to provide active localization. It is well accepted that the binding affinity, stability, and size of the ligand play a critical role for successful targeting. The conjugation of multiple antibodies to each nanocarrier enhances their avidity, and nanocarriers can be surface functionalized with multiple distinct antibodies to overcome tumor heterogeneity. Peptides and antibody fragments have been developed to overcome some of the shortcomings of antibodies, and several examples of these ligands are now under clinical development. Functional, single-domain heavy chain antibodies, referred to as nanobodies, have been raised against cancer targets which either antagonize receptor function or deliver an enzyme for prodrug activation [128]. Affibodies against a variety of cancer-related targets have been developed and are now commercially available, including: EGFR, HER2, and transferrin[27].

Cancer stem cells are characterized to be a quiescent and small cell subpopulation with different surface markers than bulk differentiated tumor cells, and to present well-developed drug resistance. While there is considerable emphasis on specific cell markers in the current cancer targeting paradigm, one must also recognize cases where cell-nonspecific approaches may be necessary for more effective and consistent therapeutic output [129].

For more sophisticated applications, where greater dosage and/or actively controlled time-release is needed, MEMS/NEMS-based devices are envisioned that can incorporate both local sensing and mechanisms to dispense drugs – see section IIIc3.

Research issues and opportunities include:

- Improvements to drug storage in the nanocarrier, including increased loading capacity.
- The mechanism(s) that drives the nanocarrier toward the target.
- Approaches to surmount bio-barriers, including mechanistic understanding [130].
- Knowledge of and control over the excretion modes.
- External/internal trigger events – ultrasound, NIR, RF, pH, ...– for drug release and/or intracellular penetration.
- Bioresponsive and self-regulated delivery systems.
- Improved knowledge of pertinent biomarkers.
- Novel approaches to accelerate the discovery process for multifunction nanocarriers, such as synthesis and automated screening.
- Reduce batch-to-batch variability by prefunctionalized biomaterials for the self-assembly of nanoparticles (NPs)[113].
- Defining the optimal interplay of biophysicochemical parameters that simultaneously confer molecular targeting, immune evasion, and drug release.

IIIb3. Innovative Therapy

The possibilities for innovative therapies are limitless. As one example, hyperthermia has been explored as a treatment for cancer for a couple of decades, but with limited success. Nanostructure-enabled innovation has rejuvenated hyperthermia with the nanoparticles extracting energy from near-infrared (NIR) [131, 132] or radio frequency (RF) [53, 133] electromagnetic fields. Lack of knowledge on the fundamental mechanisms involved has slowed the implementation of clinical protocols. For instance, recent efforts[26] for elucidating the mechanisms have demonstrated that cell membrane and cytoskelton are important loci of cell damage by both ionizing radiation and hyperthermia. Technical difficulties in developing magnetic field applicators at the frequencies and field values with concurrent compliance of the safety regulations demanded for clinical use will need to be addressed.

As a second example, photodynamic therapy is also an innovative, evolving approach for treating neovascular diseases of the eye where nanostructures can play an important role [134]; two-photon infrared, nanoplatform phototoxicity has been demonstrated for rat glioma cells [135]. A third example is magnetic nanoparticles to remove toxins from the blood; however a magnetic separator suitable for real-time clearing of magnetic nanospheres needs to be improved [136]. As a final example, self-assembling nanofibers have been shown to promote neural healing after spinal cord injury[61].

Research issues and opportunities include:

1. Develop and sustain a sufficiently robust chemistry, physics, and engineering research discovery base that is effectively coupled with the biology and medicine research community so that new opportunities are recognized and quickly exploited.
2. The paucity of information on the physical/biochemical mechanisms involved in thermosensitization, including models that can describe on a microscopic basis the interplay between physical and biochemical cell mechanisms involved.
3. For magnetic nanoparticles there are different effects to be considered for power losses in physiological conditions: a) magnetic losses through domain wall displacements (in multi-domain particles), Neel relaxation (in single domain particles); and energy loss from mechanical rotation of the particles acting against viscous forces on the liquid medium [26].

IIIc. Implants and Tissue Regeneration

IIIc1. Tissue Engineering

Tissue engineering, or regenerative medicine, is an interdisciplinary field that merges principles and innovations from the physical and chemical sciences, engineering, and the life sciences. The focus on the improvement, repair, or replacement of tissue and organ function [137-140]. The ultimate goal is to enable the body to heal itself by introducing an engineered scaffold – i.e., substitute extracellular matrix (ECM) - that the body recognizes as "self." Signals are transmitted between the cell and the ECM enabling communication for cell adhesion, migration, growth, differentiation, programmed cell death, modulation of cytokine and growth factor activity, and activations of intracellular signaling. Any scaffold material must be able to interact with cells in three dimensions and facilitate communication. Scaffold pore size, pore orientation, fiber structure, and fiber diameter are known to regulate proliferation, cellular organization, and subsequent tissue morphogenesis.

Current research in tissue engineering is approaching a major breakthrough in the treatment of injury and disease due to the ability to routinely create extracellular-matrix analogous nanofibers [137]. For example, reports of nanostructure approaches to tissue engineering have dramatically increased in the literature in the last four years (from 32 in 2004 to 219 in 2008, as identified by a literature search of ISI Web of Science using the simple keywords nanofib* and tissue). An extracellular matrix mimic must:

- Be biocompatible and function without interrupting other physiological processes.
- Not promote or initiate any adverse tissue reaction.
- Be produced by simple techniques yet versatile enough to produce a wide array of configurations to accommodate the size, shape, strength, and other intricacies of the target tissue/organ.
- Be removed via degradation of adsorption or incorporated via innate remodeling mechanisms, leaving behind only native tissues.

There are several approaches being explored for the manufacture of nanofibers for ECM such as: electrospinning [112, 141-144], phase separation [145], self-assembly [146-148], and template [149]. Each approach is different and results in a unique set of characteristics as a scaffolding system. Electrospinning is a process that can be used to fabricate scaffolds economically at large scales and can incorporate solid nanomaterials within electrospun fibers in a well-dispersed and spatially controlled manner [150]. Electrospun collagen nanofibers, for example, have been shown to produce skin substitutes with similar cellular organization, proliferation, and maturation to the current, clinically utilized model, and were shown to reduce wound contractions which may lead to reduced morbidity in patient outcomes[141]. Phase separation has generated fiber diameters in the same range as the ECM and allows for the design of macroporous structures [151]. Self-assembling peptides (SAPs) have emerged as an attractive class of 3D scaffolding materials, mainly due to the nano-scale fibrous and porous topographies that mimic the natural ECM features. Cell behavior can be controlled by cell-materials interactions if biofunctional sites are synthesized into the scaffolds [148, 152-154].

Research issues and opportunities include:

- Design nanoscale materials with functional domains that promote self-assembly into higher order scaffolds that have mechanical strength, resilience and compliance of natural ECM, while maintaining porosity and high surface area, and cues to bind circulating stem cells and then induce proliferation [154].
- Modify nanofiber properties for drug/cell recognition through the incorporation of nanoparticles and/or functionalization.
- Understand the role of nanoscale surface topography and chemistry in cell mediation through biomolecular interactions.

- Characterize the complex three-dimensional organization of the structural and functional molecules constituting the ECM .
- Incorporation of drug and gene delivery systems into biomaterial scaffolds.
- Develop and understand methods of stem cell delivery in biomaterial scaffolds overcoming the problems of cell survival.
- Biodegradable biomaterials where the by-products are bioactive agents.
- The role and efficacy of nanostructures in central nervous system regeneration.

IIIc2. Orthopedic Implants

Self-assembly and biomineralization are used in biological systems for the fabrication of many composite materials. Bone tissue is a particularly complex biological system because it contains multiple levels of hierarchical organization. Bone has been hard to replicate, so alternative materials such as Ti alloys or composites with micro-fillers have been substituted.

The lifetime of orthopedic implants is limited primarily by implant loosening, a result of interfacial breakdown and stress[62]. Implant materials - tita nium, stainless steel and cobalt alloys - are much stiffer than bone; a cement must buffer their respective mechanical properties. Polymethyl methacrylate (PMMA) is commonly used in orthopedic implant cements; without it, metal directly contacting bone leads to strong inflammatory response, and creates highly localized stresses and micro-motion of the implants. Disadvantages of PMMA include limited radiopacity, exothermic setting, and poor ossification with juxtaposed bone. The application of nanoparticles in PMMA cements, an approach that can address all of these issues, is still in its infancy. Current approaches to implant materials include creating porous surfaces [63, 155] in attempts to improve fixation, but this does not necessarily solve stiffness mismatch. Additionally, osteoblast activity can be significantly enhanced using controlled nanotopographies [156]; for instance, nanotubular titania surfaces have been shown to provide a favorable template for bone cell growth and differentiation [157, 158]. There is a published broad review on the topic of nanomaterial interactions with proteins and cells [159].

In a somewhat simpler application, new dental restorative materials already in the marketplace are exploiting composites incorporating nanoparticles of silica (for improved mechanical properties and luster) and zirconia (for improved radiopacity) in a polymer matrix [160].

Research issues and opportunities include:

- Dispersion of nanoparticles evenly in the bone cement matrix.
- Inadequate knowledge of how to engineered surfaces with nanoscale features to affect vascular and bone cell adhesion, thereby providing a bioactive surface for bone integration.

IIIc3. Implanted drug dispenser / factory

The next generation of drug delivery systems, in addition to having spatial and temporal control, is expected to be "smart" and to enable therapy that is responsive to the patient's specific needs. These advanced systems would protect drugs from environmental or biological degradation in the body, use closed-loop control to assist the patient with homeostasis, and provide autonomous drug administration. MEMS/NEMS methods can provide a sophisticated approach. With controlled delivery, appropriate and effective amounts of drug might be precisely calculated by the controller and released or manufactured at an appropriate time. Present MEMS based microfluidic drug delivery devices [161, 162] include: microneedle-based transdermal devices, osmosis-based devices, micropump-based devices, and microreservoir-based devices. Micropumps for transdermal insulin delivery, injection of glucose for diabetes, and administration of neurotransmitters to neurons have been reported [163].

The fabrication of nano- and micro-scale 3D programmable volume enclosures (voxels) to encapsulate nano-scale quantities of various materials is expected to greatly expand current capabilities. If cells/tissue are incorporated into the voxel as the drug manufacturing mechanism, the enclosure walls must have pores small enough to prevent immunoresponse, but large enough to permit the suffusion of metabolites [57]. Nanoscale approaches to power for such devices include: stored energy (battery) [164], wireless transfer [165, 166], and local generation [167-170].

Research issues and opportunities include:

- Biocompatibility of the implanted devices.
- Implantable power sources – battery, wireless, scavenging.
- Devices with low power, low heat dissipation and high sensitivity.
- Technologies for the development of new generations of synthetic polymers that can change their molecular conformation in response to changes in external stimuli (mechanical, temperature, pH, etc.).
- The use of low-cost technologies such as injection molding or low-cost substrates such as PDMS (polydimethylsiloxane) or polyimide for the fabrication of microfluidic devices.
- Sensor technology for the assessment of the interface activity and the progress of implant integration and functional state.
- Development of effective, long-lived, implanted systems incorporating transplanted living cells for the production of needed chemicals.
- Biomimetic membranes with built-in functionality, which can mimic real cell membranes for (stem) cell attachment and/or stimulation (proliferation, differentiation).

IIIc4. Implants interacting with the Central Nervous System (CNS)

The nervous system has a poor healing capacity. Additionally, an aging population leads to more persons acquiring disabilities - such as hearing loss, stroke, and Parkinson's disease. The demand for solutions is growing [171]. The meeting report [172] "Smart Prosthetics: Exploring Assistive Devices for the Body and Mind" focused on several themes relevant to future prosthetics where it is expected that the nanoscale will be important. The potential of nanotechnology applications in neuroscience is becoming accepted and is the subject of several reviews [69, 110, 173-175].

Dating from 1972, about 100,000 patients worldwide have received cochlear implants. The current state of this technology is bulky, difficult for the surgeon to implant and doesn't allow a broad range of perceived frequencies [176]. The human auditory nerve contains ~30,000 axons, which cochlear implants stimulate currently with 3-22 electrodes. MEMS, micromechanical devices are being developed to ameliorate these problems. Since the human ear itself already uses nanostructures [177], continued miniaturization beyond the microscale is certain to provide additional improvements.

In contrast to the auditory nerve, the optic nerve has about a million fibers. Visual prosthesis must also deal with two dimensional spatial data and the highly complex signal processing that occurs in the retina before transmission to the brain. A fully implantable retinal prosthesis would ideally capture all of the functions performed by the mammalian retina in one autonomous device. It is postulated that the needed computations can be performed at an energy efficient and physical scale comparable to biology by incorporating principles derived from neural circuits into nanoelectronic circuits [178-180].

For the control of artificial limbs the next generation of prosthetics will use regions of undamaged nervous tissue to provide command/sensory signals [181]. However, problems range from improper neuronal adhesion to inadequate signal stability. Implanted electrodes do not remain statically placed due to different flexibilities in implants materials versus tissue, or the growth of fibrous tissue around the implant. So new materials solutions are needed [182, 183]. Single wall carbon nanotubes (SWCNT) have received promising attention because of their unique physical and chemical features [184, 185]. Nanostructured porous silica is found to be more biocompatible than a smooth surface, producing less glial activation and allowing more neurons to remain close to the device [186]. Light-activated semiconducting nanoparticles have been shown to wirelessly stimulate neurons in the rat brain [187].

Research issues and opportunities include:

- Fundamental studies to find highly stable substrate and electrode materials, reliable and robust assembly, and encapsulation materials to deliver vision implants with lifetimes, biocompatibility and functionalities that are comparable to cardiac pacemakers and cochlear implants [188].
- Improving how the implanted device responds to stimuli in its local mechanical environment.
- Improving the transfer of information between the brain and/or nervous system and the device.

IIId. Biological Systems Engineering

A goal of systems biology is to fundamentally transform the practice of medicine [189-193]. Systems biology is the study of an organism, viewed as an integrated and interacting network of genes, proteins and biochemical reactions. The study of systems biology has been aided by cyber-enabled information storage/processing, advances in nanotechnology, advances in modeling and simulation, and the infusion of science and scientists from other disciplines, e.g. computer scientists, mathematicians, physicists, and engineers.

Enabled in part by the rapid progress in nanoscale science and engineering, and the growing sophistication of computers and cyberinfrastructure, systems biology is a field coming of age. The potential impact on medicine and health is enormous. Nanoscience can accelerate this field in a number of ways. First, it provides the ability to examine the properties of individual nanostructures rather than the average properties measured by techniques that require ensembles for adequate signal to noise. Work at the nanoscale enables the study of single molecule properties previously very hard to measure, such as protein folding/unfolding [194, 195], molecular motors [194], and DNA/RNA sequencing [196-199]. Second, as microfluidics and sensing technologies become further miniaturized, there will be growing capability to provide arrays that can potentially detect and identify many constituents in a biological sample in time frames of minutes rather than hours or days. Two microfluidic foundries are now available for the academic community [200]. Third, somewhat incidental but still important, the advent of nanoelectronics is continuing the increase in computational power that will be essential to model a system as complex as a cell.

The complexity of biological systems will continue to require the sampling of multiple cells. However, as with single molecule studies, the capability to probe individual cell behavior is essential to rapid progress. Microfluidics offers analytical devices with length scales that are: a) comparable to the intrinsic dimensions of prokaryotic and eukaryotic cells, organelles and, b) the length scale of diffusion of oxygen and carbon dioxide in tissues [108, 200, 201]. The growing availability and sophistication of microfluidic chips will accelerate single cell studies [202-204]. As examples of new capabilities, nano-enabled probes have been shown to physically penetrate the cell membrane with minimal disruption [205], improve the resolution of optical probes with 3D resolution at the nanoscale [206], acquire spectroscopic [73] and fluorescent signatures [72, 207], actuate membrane receptor mediated signal transduction [208], probe cell mechanical behaviors [98, 209, 210], characterize calcium release [211], grow and probe neurons [212, 213], and probe single cell motility and metabolic calorimetry [214].

In addition to the controlled study of single cells previously mentioned, microfluidics is being utilized to study processes such as blood clotting [215]. Scaling (thousands of identical microfluidic structures) is of increasing importance in biology as the field moves toward quantitative data because it allows multiple parallel experiments under identical conditions[202].

Research issues and opportunities include:

- Microfluidic structures utilizing biocompatible materials.
- Improvements in pumping and valving.
- Improvements in on-chip sensitivity (excitation/detection) to permit single molecule detection in biological media, including inside a cell.
- Detailed understanding of single macromolecular folding/unfolding events and the role of chaperone molecules.
- Technologies that include electronic and/or communication components in forms of nanowires and nanopores for the stimulation and biosensing of cells.

IIIe. Innovations in Medical Instrumentation and Devices

Work at the nanoscale requires the continued miniaturization of measurement devices, both for spatial localization and for augmented sensitivity to the small signals associated with a nanostructure. Adaptations of these new devices for medical applications are forthcoming. As examples, the force microscope is capable of measuring differences between cancer and normal cells[98] and bone viability [216, 217]. Carbon nanotubes (CNTs) function better than glass pipettes for cellular delivery [218]. The incorporation of micro-nano-devices into catheters and other instruments is growing [219], including incorporation of Ag nanoparticles to impart antimicrobial activity [220]. Nanostructures are enabling electronic circuitry on flexible substrates, including high performance circuit elements (e.g., Si or CNT devices) [221, 222]. One can envision the incorporation of signal processing and sensing capabilities into mechanically flexible implants and even surgical gloves that might detect important parameters [75]. Nanopore-based devices show considerable promise for low cost, rapid DNA sequencing [223].

IV. Present Federal Programs

There are several U.S. Federal agencies that fund pertinent health research. The foremost is the National Institutes of Health. In addition, NASA is interested in medical practice in space; DOD has an interest in warfighter health issues and battlefield medicine; NSF provides the foundations of medicine in systems biology; EPA is concerned with impact on living systems in the environment; and USDA is concerned with impact on agriculture. The total Federal investment in the National Nanotechnology Initiative is given in Table 1. Table 3 provides an estimate of the investment more directly relevant to medicine and health.

IVa. National Institutes of Health (NIH)

There is an individual investigator-initiated research investment in nano-enabled medicine of about $140M/yr distributed throughout the NIH; an additional $50M is invested in centers. The NIH investment in nano-enabled medicine is monitored by a Trans-NIH Task Force.

Nanomedicine Initiative

The NIH has a Nanomedicine Implementation Group with membership from the various institutes [6]. Under its Roadmap for Medical Research – New Pathways to Discovery, NIH has established a national network of eight Nanomedicine Development Centers (NDC, see Table 4), which serve as the intellectual and technological centerpiece of the Nanomedicine initiative. These collaborative centers are staffed by multidisciplinary research teams comprising biologists, physicians, mathematicians, engineers and computer scientists. In the initial phase of the program (FY2005-FY2010), research has been primarily directed toward gathering extensive information about the chemical and physical properties of nanoscale biological structures. A second phase for the program has been approved during which the acquired fundamental knowledge and developed tools will be applied to understanding and treating disease. The Centers reach out to clinical investigators with ongoing opportunities for potential medical applications that build on the science emerging from the NDC.

National Institute of Biomedical Imaging and Bioenginering (NIBIB)

Unlike any other NIH Institute or Center, the National Institute of Biomedical Imaging and Bioengineering's mission is focused on emerging technology development. The Institute has a mandate to enable and promote fundamental discoveries, and to support the design, development, translation, and assessment of technological capabilities in biomedical imaging and bioengineering. NIBIB has programs in Micro/Nano systems [224] and Nanotechnology [225]. In addition, NIBIB sponsors Centers that are pertinent to nanotechnology. The pertinent Biotechnology Resource Centers include the Biomicroelectromechanical Systems (BioMEMS, Mass General Hospital), Biophysical Imaging Opto-Electronics (Cornell), National ESCA and Surface Analysis Center (Univ. Wash), Tissue Engineering (Tufts Univ.), and Computer Integrated System for Microscopy and Manipulation (Univ. North Carolina). The pertinent Point-of-Care (POC) Technologies Research Network includes centers on Emerging Neurotechnologies (Univ. Cincinnati), Rapid Multipathogen Detection (UC Davis), Diagnostics for Global Health (PATH, Seattle), and Sexually Transmitted Diseases (Johns Hopkins Univ.). These centers coordinate development, clinical evaluation, and reduction to practice of new POC devices. NIBIB sponsors an Interfaces Initiative for Interdisciplinary Graduate Research Training (T32) program with $3-4M/yr devoted to nanotechnology, and works with NSF and Howard Hughes Medical to address interdisciplinary training.

National Cancer Institute (NCI)

Initiated in 2004, the NCI Alliance for Nanotechnology in Cancer encompasses four major program components: Centers for Cancer Nanotechnology Excellence (CCNE, see Table 4), the Nanotechnology Characterization Laboratory (NCL, in collaboration with FDA and NIST), the Cancer Nanotechnology Platform Partnerships (CNPP, see Table 4), and a multidisciplinary research training and team development program. The funding level for the Alliance is projected at $144M over five years [226]. The partnerships are designed to develop technologies for new products in six key partnership areas: molecular imaging and early detection, *in vivo* imaging, reporters of efficacy (e.g., real-time assessments of treatment), multifunctional therapeutics, prevention and control, and research enablers (opening new pathways for research). NCI and NSF are collaborating in training programs for U.S. science and engineering doctoral students through the Integrative Graduate Education and Research Traineeship Program (IGERT) – Rutgers (nanopharmaceutical), Northeastern (brain-machine), University of New Mexico (micro-nano-bio interfaces) and University of Washington (nanotechnology workforce).

Table 3. Approximate Federal Investment in Nano-Science/Engn Relevant to Medicine/Health

Agency		FY08
NIH	NCI - Alliance for Nanotechnology (largely centers)	30M
	NHLBI – Centers of Excellence in Nanotechnology	10M
	Nanomedicine Centers	10M
	Other	140M
NSF	Chemical, Biological, Environmental, and Transport Systems Div	25M
	Biological Sciences Directorate	25M
DOD	Mostly Multidisciplinary Univ. Research Initiative (MURI) efforts	5M

Table 4. Federal Multidisciplinary Center Programs Relevant to Medicine/Health

Program	PI Name	Institution Name	Center Title
NIH			
CCNE	Rudolph Juliano	Univ. of North Carolina	Carolina Center of Cancer Nanotechnology Excellence
CCNE	Sanjiv Gambhir	Stanford Univ.	Ctr for Cancer Nanotechnology Excelle-nce Focused on Therapy Response
CCNE	Robert Langer	MIT	Center of Cancer Nanotechnology Excellence
CCNE	Sadik Esener	UC, San Diego	Ctr of Nanotechnology for Treatment, Understanding, & Monitoring of Cancer
CCNE	Shuming Nie	Georgia Inst of Technol.	Nanotechnology Center for Personalized and Predictive Oncology
CCNE	Chad Mirkin	Northwestern University	Nanomaterials for Cancer Diagnostics and Therapeutics
CCNE	James Heath	California Inst of Technol.	Nanosystems Biology Cancer Center (NSBCC)
CCNE	Samuel Wickline	Washington Univ.	The Siteman Center of Cancer Nanotechnology Excellence
CNPP	Douglas Hanahan	UC, San Francisco	Detecting Cancer Early with Targeted Nano-probes for Vascular Signatures
CNPP	James Baker	Univ. Michigan	DNA-Linked Dendrimer NP Systems for Cancer Diagnosis & Treatment
CNPP	Kattesh Katti	Univ. of Missouri	Hybrid nanoparticles in Imaging and Therapy of Prostate Cancer
CNPP	Scott Manalis	MIT	Integrated System for Cancer Biomarker Detection
CNPP	Panos Fatouros	VCU	Metallofullerene Nanoplatform for Imaging & Treating Infiltrative Tumor
CNPP	Paras Prasad	SUNY, Buffalo	Multifunctional Nanoparticles in Diagnosis & Therapy of Pancreatic Cancer
CNPP	Miqin Zhang	Univ. Washington	Nanotechnology Platform for Pediatric Brain Cancer Imaging and Therapy
CNPP	Jan Schnitzer	Sidney Kimmel Cancer Ctr	Nanotechnology Platform for Targeting Solid Tumors
CNPP	Mansoor Amiji	Northeastern Univ.	Nanotherapeutic Strategy for Multidrug Resistant Tumors
CNPP	Chun Li	UT Anderson Cancer Ctr	Near-infrared Fluorescence NP for Targeted Optical Imaging
CNPP	Ravindra Pandey	Roswell Cancer Inst.	Cancer Nanotechnology Platforms for Photodynamic Therapy & Imaging
CNPP	Tayyaba Hasan	Mass General Hospital	Photodestruction of Ovarian Cancer: EfbB3 Targeted Aptamer-NP
PEN	Karen Wooley	Washington Univ.	Integrated Nanosystems for Diagnosis and Therapy

Table 4. (Continued)

Program	PI Name	Institution Name	Center Title
PEN	Gang Bao	Georgia Inst of Technol.	Nanotechnology: Detection & Analysis of Plaque Formation
PEN	Jeffrey Smith	Burnham Institute	Nanotherapy for Vulnerable Plaque
PEN	Ralph Weissleder	Mass General Hospital	Translational Program of Excellence in Nanotechnology
NDC	Wah Chiu	Baylor College of Med.	Center for Protein Folding Machinery
NDC	Chih-Ming Ho	UC, Los Angeles	Center of Cell Control
NDC	Wendell Lim	UC, San Francisco	Engineering Cellular Control: Synthetic Signaling and Motility Systems
NDC	Gang Bao	Georgia Inst. of Technol.	Nanomedicine Center for Nucleoprotein Machines
NDC	Michael Sheetz	Columbia Univ.	Nanotechnology Center for Mechanics in Regenerative Medicine
NDC	Eric Jakobsson	UIUC	National Center for Design of Biomimetic Nanoconductors
NDC	Ehud Isacoff	UC, Berkeley	Optical Control of Biological Function
NDC	Peixuan Guo	Univ. of Cincinnati	Phi29 DNA-Packaging Motor for Nanomedicine
NSF			
NSEC	Dawn Bonnell	Univ. Pennsylvania	Center for Molecular Function at the Nanoscale
NSEC	Vicki Colvin	Rice Univ.	Center for Biological and Environmental Nanotechnology
NSEC	Richard Siegel	RPI	Center for Directed Assembly of Nanostructures
STC	Harold Craighead	Cornell Univ.	The Nanobiotechnology Center
MRSEC	Mehmet Sarikaya	Univ. of Washington	Genetically Engineered Materials Science and Engineering Center
DOD			
MURI	Jimmie Xu	Brown Univ.	Direct Nanoscale Conversion of Biomolecular Signals
MURI	G. Oberdorster	Univ. of Rochester	Physicochemical Characterization & Toxicology for Air/Space
MURI	Naomi Halas	Rice Univ.	Nanoscale Optical Imaging with Integrated Spectroscopies
MURI	H. Abarbanel	UCSD	Chem. Discrimination & Localization Using Bio-Based Olfactory Processing
MURI	Chad Mirkin	Northwestern Univ.	Bioinspired Supramolecular Enzymatic Systems

National Heart, Lung and Blood Institute (NHLBI)

Starting in 2004, the National Heart, Lung, and Blood Institute (NHLBI) of the National Institutes of Health (NIH) has made four 5-year awards to initiate a unique, diverse, and nationwide Program of Excellence in Nanotechnology [227] (PEN, see Table 4). This program brings together bioengineers, materials scientists, biologists, and physicians who work in interdisciplinary teams to spur the development of novel technologies to diagnose and treat heart, lung, and blood diseases.

National Institute of Environmental Health Sciences (NIEHS) / National Toxicology Program (NTP)

The National Institute of Environmental Health Sciences administers the National Toxicology Program which has research activities focusing on 4 classes of nanostructured materials [228]:

- Metal oxides: the initial focus is on nanoscale titanium dioxide and zinc oxide due to their presence in cosmetics.
- Fluorescent crystalline semiconductors (quantum dots): the initial focus is on cadmium selenide/zinc sulfide spheres and rods of varying sizes and surface chemistry as a model system.
- Fullerenes: the initial focus is on carbon based fullerenes of varying cage size and surface derivatisation.
- Nanotubes: the initial focus is on single walled carbon nanotubes. Through a NIEHS-NIOSH (National Institute for Occupational Safety and Health) interagency agreement the NTP is supporting the development of exposure systems for inhalation toxicity studies of single walled nanotubes.

National Institute of General Medical Sciences (NIGMS)

The National Institute of General Medical Sciences (NIGMS) has research on, and development of, new and improved instruments, methods and technologies for nanoscience, and for the analysis of single protein and nucleic acid molecules and their complexes both *in vivo* and *in vitro.*

National Center for Research Resources (NCRR)

The National Center for Research Resources (NCRR) consortium is to transform how clinical and translational research is conducted, ultimately enabling researchers to provide new treatments more efficiently and quickly to patients [229]. Clinical and Translational Science Awards (CTSA) support clinical and translational research by providing access to clinical and translational research resources developed by the CTSAs, government sponsored research communities, government agencies or private sector. The NCRR consortium could become a powerful tool in the translation on nano-science/engineering discoveries.

NanoHealth Enterprise, public private partnerships

NIH is exploring the NanoHealth Enterprise which would comprise an integrated, interdisciplinary program that draws upon the expertise and interests of the NIH institutes and centers, in partnership with private industry, to address critical research needs for the safe development of nanoscale materials and devices. This initiative outlines an integrated, interdisciplinary program that draws upon the expertise and interests of the NIH institutes and centers, and addresses critical research needs for the safe development of nanoscale materials and devices. The initiative proposes five components: Materials Science Research, Basic Biology Research, Pathobiology Research, Informatics, and Training.

Small Business Innovative Research (SBIR) / Small Business Technology Transfer Research (STTR)

NIH is one of several agencies that have a "nano" specific topic in its SBIR/STTR announcement. Few small businesses possess the highly specialized resources needed for nanoengineering, so applications are encouraged from teams of investigators from commercial, academic and other sectors of the research community. The NIH Pipeline to Partnerships is a virtual space for NIH SBIR/STTR awardees to showcase technology and product development for an audience of potential strategic partners and investors.

IVb. National Science Foundation (NSF)

While much of the extensive (~$350M) NSF investment in "nano" has the potential to impact medicine and health, the most directly involved programs are located in the Biological Sciences Directorate and in the Chemical, Biological, Environmental and Transport Systems Division of the Engineering Directorate. Biological Sciences has the role to promote and advance scientific progress in biology. The Chemical, Bioengineering, Environmental and Transport Systems Division (CBET) supports research in bioengineering (among other topics). These two programs provide much of the underlying science and engineering base for medicine and health applications, which is critical to rapid advancement.

The CBET Division has programs in "Integration of Life Sciences with Engineering", as well as "Nanoscale Science and Engineering." The current high-emphasis research and education areas include: post-genomic engineering, tissue engineering, biophotonics, and nano-biosystems. CBET has approximately $25M invested in nano-bio projects.

The Molecular and Cellular Biosciences (MCB) Division effort under the Biological Sciences Directorate at NSF emphasizes systems biology; it has approximately $25M invested in nano-bio projects. The research includes databases and informatics, instrument development, biomolecular systems, cellular systems, and genes and genome systems. The latter three encourage multi-disciplinary approaches, including research carried out at the interfaces of biology, physics, chemistry, mathematics and computer science, and engineering.

The National Nanotechnology Infrastructure Network (NNIN) is an integrated partnership of thirteen user facilities, supported by NSF, providing unparalleled opportunities for nanoscience and nanotechnology research [230]. The network provides extensive support in nanoscale fabrication, synthesis, characterization, modeling, design, computation and hands-on training in an open, hands-on environment, available to all qualified users.

IVc. Food and Drug Administration (FDA)

The regulation of nano-enabled products may involve more than one traditional FDA category, for example a "drug" delivery "device". In these cases the assignment of regulatory lead is the responsibility of the Office of Combination Products. To facilitate the regulation of nanotechnology products, the Agency has formed a NanoTechnology Interest Group (NTIG), comprised of representatives from all its Centers. The NTIG meets quarterly to ensure there is effective communication between the Centers. A FDA task force report on nanotechnology is available [21]. The FDA is a cosponsor of the Nanotechnology Characterization Laboratory, along the NCI and NIST, and the nanostructure evaluation in the National Toxicology Program with NIEHS. There is a FDA intramural research program, but it does not presently have any "nano" focused projects.

The FDA and Alliance for NanoHealth co-convened a workshop on nanomedical regulatory science in Houston in March 2008 to identify the top scientific hurdles in bringing nanoengineered products to patients, specifically in the pre-clinical, clinical and manufacturing phases of product development [231]. Six priority areas were identified:

- Determination of the distribution of nanoparticulate carriers in the body following systemic administration through any route.
- Development of imaging modalities for visualizing the biodistribution over time.
- Understanding mass transport across compartmental boundaries in the body.
- Develop new mathematical and computer models that will lead to predicting risk and benefit parameters.
- Establish standards or reference materials and consensus testing protocols that can provide benchmarks for the development of novel classes of materials.
- Realization of an analytical toolkit for nanopharmaceutical manufacturing, accompanied by specification sheet of toxicological, safety, and biodistribution properties obtained through standardized, validated methods.

IVd. Department of Defense (DOD)

The DOD does not have any appreciable program in nanomedicine *per se*. The DARPA Defense Science Office has thrusts on tactical and restorative biomedical technologies that may exploit nanotechnologies. There are some limited efforts from the various service research offices are examining how to exploit nanotechnology with medical implications; five such Multidisciplinary University Research Initiatives (MURIs) are listed in Table 4 and ONR has a recent initiative on Autonomous Devices for Advanced Personnel Treatment. The Army funds the Institute of Soldier Nanotechnologies (MIT University Affiliated Research Center) that addresses some medical applications. The Army Telemedicine and Advanced Technology Research Center (TATRC) oversees a diverse portfolio, largely Congressional adds to the DOD budget, ranging from new nanomaterial-based contrast agents for cardiac and brain imaging to novel drug delivery systems for the treatment of cancer. In each case TATRC assists the program in identifying military needs, defining metrics, and comparing the new technology to existing methods.

IVe. Department of Energy (DOE)

The DOE Nano Centers are user facilities for interdisciplinary research at the nanoscale. Each of the five Centers is co-located with other large scientific facilities to take advantage of complementary capabilities, such as the Spallation Neutron Source at Oak Ridge, the synchrotron light sources at Argonne, Brookhaven and Lawrence Berkeley, and semiconductor, microelectronics and combustion research facilities at Sandia and Los Alamos. The Centers contain clean rooms, laboratories for nanofabrication, one-of-a-kind signature instruments, and other instruments (such as nanopatterning tools and research-grade probe microscopes) not generally available except at major scientific user facilities.

IVf. European Union (EU)

The Framework 7 included an ERA-NET (European Research Area – network) on nanomedicine project - NMP-2008-4.0-13 - with the expectation of: improved coordination and reduced overlapping and fragmentation; achieving critical mass and ensuring better use of limited resources; sharing good practices in implementing research programs; and promoting transnational collaborations and generate new knowledge. In FY08 approximately €8M was allocated for the first call ERA-NET plus (of which nanomedicine is a part).

As another example, the project "Healthy Aims: Developing New Medical Implants and Diagnostic Equipment" is a € 23M, four year project to develop intelligent medical implants and diagnostic systems. While not constrained to nanotechnology-enabled, products under development will almost certainly benefit from nanoscale capabilities. The funded projects include: retinal implant, functional electrical stimulation of systems for restoration of upper-limb movement as well as bladder and bowel control, cochlear implant, glaucoma sensor, intracranial pressure sensor, and a sphincter sensor for monitoring bladder pressure [232].

The European Technology Platform: Nanomedicine – Nanotechnology for Health identifies the following as strategic research priorities [7]:

Diagnostic issues

- *In vitro* diagnostics.
- *In vitro* and *in vivo* imaging.

Drug/Delivery/Therapeutic Issues

- Improving targeting agents.
- Formulation and stability of pharmaceutics.
- Easier routes of administration – crossing biological barriers.
- Nanodevices for targeted delivery.
- Bioactive signaling molecules.
- Cell-based therapies.

Implant/Tissue Regeneration Issues

- Interactions between biological systems and artificial nanostructures.
- Intelligent biomaterials and smart implants.

Overarching Issues

- Basic science deficiencies.
- Medical devices.
- Moving established and novel nano-therapeutic delivery systems from the laboratory to the clinic.

V. Recommendations

Research opportunities and challenges have been identified in each of the subcategories in Section III; they are many. It should be noted that more generic questions were addressed in this workshop, and that the research funding levels and prioritization amongst these opportunities and challenges was beyond its scope. From the discussions at the Re-Engineering Basic and Clinical Research to Catalyze Translational Nanoscience Workshop at USC, augmented by literature search and subsequent evaluation by other experts, the following overarching recommendations are made:

- Medicine enabled by nanoscale science and engineering
    - The continuing progress in nanoscale science and engineering promises to create revolutionary opportunities for medicine and health; the investment in the basic discoveries should not be diminished. Rather additional funding should be found for the translational efforts.
    - The Trans-NIH Nano Task Force deserves kudos for its efforts to inject nanoscience into the NIH portfolio, but only a handful of the NIH Institutes (NCI, NHLBI, NIBIB) have created explicit programs to exploit the nanoscale. As progress at the nanoscale continues to progress, other NIH Institutes should be encouraged to develop explicit efforts to engage the nanoscience and nanoengineering communities.
    - The NIBIB website, Nanotechnology at NIH, provides a central location for the various NIH programs, is a valuable resource for the science/engineering communities, and should be kept up to date.
    - As nano-enabled improvements are incorporated into functional medical devices and systems, it will become more difficult to track the "nano" impact. NIH is encouraged to make that effort, both to better understand where nanotechnology provides viable solutions, and to document those contributions for inclusion in social and political debates.
    - Continue and expand the efforts to build bridges between the physical sciences, engineering, the medicine/health professionals, and the medical technology industries. Centers are a means to accomplish this goal, but affect only a limited

number of individuals. There are several extant professional forums that address translational nanomedicine, including the BioMaterials Society, but the cross-fertilization between clinical physicians and the nanoscience research investigators is minimal. Gordon Conference-like meetings with limited attendance and a site designed to encourage full participation over a weeks span should be encouraged. It may be necessary to offer financial assistance or continuing education credits as incentives to clinicians to enable their participation. One of the more promising outcomes from the Re-Engineering Basic and Clinical Research to Catalyze Translational Nanoscience Workshop was interaction amongst the participants with the anticipation of fruitful collaborations.

  - The NIH should encourage its employees and grantees to contribute to the much needed efforts in developing standards by the American Society of Testing and Materials (ASTM), American National Standards Institute (ANSI), and International Standards Organization (ISO). The development of good standards
  - The Nanotechnology Characterization Laboratory (initiated by NCI, NIST and FDA), or its equivalent(s), should be expanded for access by all nanomedicine research. Because of their relative newness and the difficulty in their analysis, nanostructures tend not to be well characterized. This can lead to erroneous interpretations of experimental work and has been a source of problems.

- Translation
  - The Clinical and Translational Science Awards (CTSA) program should explicitly encourage injection of nano-enabled technology into clinical settings. The National Center for Research Resources (NCRR) consortium is meant to transform how clinical and translational research is conducted, ultimately enabling researchers to provide new treatments more efficiently and quickly to patients. Nano-enabled medicine and health technologies will be rapidly maturing; facilitating their translation into the clinic will be highly worthwhile.
  - The Bioengineering Nanotechnology SBIR/STTR announcements provide a useful approach to translation. With due attention to return on investment, continuing these explicit SBIR/STTR announcements is encouraged.
  - There should be a translation program identified and publicized for nano-enabled medicine/health. Several NIH Institutes have a cooperative program in translational research. Those programs facilitate solicitation, development, and review of therapy-directed projects to accelerate the translation of basic research discoveries into therapeutic candidates for clinical testing. Since multidisciplinary approaches are important to nanomedicine, and many of the contributors are not familiar with NIH, the NIBIB 'Nanotechnology at NIH' website should provide explicit mention of this, or similar, opportunities.
  - The NIH should explore partnering with NSF and DOE to expand the nanoscale user facility capabilities with specific focus on nanomedicine needs. While the NNI has funded a number of User Facilities for Nanoscale fabrication/ characterization, they are not focused on medical needs. Biocompatible materials and materials processing are frequently not compatible with traditional semiconductor processing. In the UK, a joint venture has been formed between

the University College of London and Imperial College BioNano Consulting to better enable industry to access the UK leading research capability in the field of bionanotechnology. It is meant to help companies with prototyping and characterization.
  - There is need for a science base to develop understanding of the critical parameters that can provide generic guidance to the FDA approval process. The FDA and Alliance for NanoHealth Workshop (see section IVc) identified six priority areas for research. Since the FDA research budget is limited, NIH (and NSF) should work with the FDA to create programs addressing those areas.
  - The Department of Commerce should work to ensure well constructed patents in the highly multidisciplinary nano-enabled biotechnology topic and to facilitate workable licensing arrangements amongst the various commercialization partners. Establish a nanomedicine group within the USPTO. Given the complexities of
  - Provide mechanisms for better interfacing between industry, academia and government. Establish protocols and technology transfer policies that foster translation of nanomedicine. Some suggestions are: a) Simplify the pathway from invention to innovation/commercialization though new IP practices. The time and expense required for negotiating collaboration and licensing agreements must be reduced; b) Encourage industry participation in NIH Nanomedicine Centers, both as advisory board members and researchers; c) Encourage industrial participation on NIH peer review panels; and d) To accelerate translation, encourage industrial participation on NIH grants, both as consultants and where appropriate as researchers.
  - NIH should explore mechanisms such as the DARPA programs and the NIST Technology Improvement Program whereby industry can participate in translation efforts. The pending NanoHealth Enterprise effort to promote public-private partnerships could have real value in accelerating translation (as well as ameliorating ESH concerns). However, care must be taken to fully engage the private sector rather than impose government priorities.
- Environmental / Safety / Health (ESH) Concerns
  - The NNI reauthorization legislation specifically identifies this topic for augmented investment. However, the size of any investment must be carefully examined to ensure that adequately characterized materials are utilized; otherwise improper conclusions may be drawn from a study. As noted above, the Nanotechnology Characterization Laboratory, should be either expanded to service all of the nanomedicine efforts or a sister laboratory created to serve that function.
  - There is a challenge to create and maintain databases that will be easily accessed by all. The NIH NanoHealth Enterprise is looking to develop public-private partnerships about three topics – nanobioinformatics, nanostructure characterization, and nanostructure/bio interactions – and is one possible approach to addressing this need.
  - The funds identified for ESH research in the NNI is growing. The ESH work will also be relevant and important to medicine/health. Conversely, research in

medicine/health will certainly involve fate and effects of nanostructures in living systems and will be useful for ESH. Looking for harmful and beneficial effects of a nanostructure are two sides of the same coin; there is need keep the two communities working closely together.
  - ESH concerns are not unique to the U.S. The NSET agencies, and NIH in particular, must be aggressive in fostering international collaborations to take advantage of other programs.
- Systems biology
  - To address the breadth and complexity of the science and engineering challenges inherent in systems biology, a concerted national program is warranted.

REFERENCES

[1] The National Nanotechnology Initiative. Research and Development Leading to a Revolution in Technology and Industry, Supplement to the President's FY2009 Budget: http://www.nano.gov/NNI_09Budget.pdf.
[2] The National Nanotechnology Initiative. FY2009 Budget and Highlights: http://www.nano.gov/NNI_FY09_budget_summary.pdf.
[3] Nanoscience and Nanotechnology; Shaping Biomedical Research, Symposium Report, June 2000: http://ww.becon.nih.gov/nanotechsympreport.pdf.
[4] Nanobiotechnolog;. Report of the National Nanotechnology Initiative Workshop, October 9-11, 2003: http://www.nano.gov/nni_nanobiotechnology_rpt.pdf.
[5] Moghimi, SM; Hunter, AC; Murray, JC. Nanomedicine: Current Status and Future Prospects. *FASEB Journal* 2005, 19, 311-330.
[6] http://nihroadmap.nih.gov/nanomedicine.
[7] European Technology Platform - Nanomedicine: Nanotechnology for Health: http://cordis.europa.eu/nanotechnology/nanomedicine.htm.
[8] www.esnam.org.
[9] www.clinam.org.
[10] Giersig, M; Khomutov, GB. *Nanomaterials for Application in Medicine and Biology. ISBN 978-1-4020-6828-7.* Springer; 2008.
[11] Jain, KK. *The Handbook of Nanomedicine. ISBN 978-1-60327-318-3.* Humana Press; 2008.
[12] www.nano.org.uk/nanomednet/.
[13] www.rsc.org/Publishing/Journals/lc/index.asp.
[14] www.nanodds.org.
[15] www.aananomed.org.
[16] www.biomaterials.
[17] www.avs.org.
[18] www.esf.org/activities/esf-conferences/.
[19] http://nm09.nanoevents.de/.
[20] Bawa, R; Johnson, S. The Ethical Dimensions of Nanomedicine.*Medical Clinics of North America* 2007, 91:881-887.

[21] Nanotechnology. A Report of the U.S. Food and Drug Administration Nanotechnology Task Force, July 25, 2007: http://www.fda.gov/nanotechnology/taskforce/report2007 .html.
[22] Strategy for Nanotechnology-Related Environmental, Health, and Safety Research: www.nano.gov/NNI_EHS_Research_Strategy.pdf.
[23] Nanomaterials. a risk to health at work? Report of the First International Symposium on Occupational Health Implications of Nanomaterials, Buxton, Derbyshire, UK, October 12-14, 2004, The Social and Economic Challenges of Nanotechnology, Economic & Social Research Council (UK): www.hsl.gov.uk/capabilities/nanosymrep _final.pdf.
[24] Towards Predicting Nano-Biointeractions. An International Asessment of Nanotechnology Environment, Health and Safety Research Needs, International Council on Nanotechnology 4 (2008): http://cohesionrice.edu/CentersAndInst/ICON /emplibrary/ICON_RNA_Report_Fu ll2.pdf.
[25] Rickerby, DG. Nanotechnological Medical Devices and Nanopharmaceuticals: The European Regulatory Framework and Research Needs.*Journal of Nanoscience and Nanotechnology,* 2007, 7, 4618-4625.
[26] Goya, GF; Grazu, V; Ibarra, MR. Magnetic Nanoparticles for Cancer Therapy.*Current Nanoscience,* 2008, 4,1-16.
[27] Alexis, F; Rhee, J-W; Richie, JP; Radovic-Moreno, AF; Langer, R; Farokhzad, OC. New Frontiers in Nanotechnology for Cancer Treatment.*Urologic Oncology* 2008, 26, 74-85.
[28] Woodrow Wilson Center Project on Emerging Nanotechnologies: http://www .nanotechproject.org/inventories/.
[29] Wagner, V; Dullaart, A; Bock, A-K; Zweck, A. The Emerging Nanomedicine Landscape. *Nature Biotechnology,* 2006, 24, 1211-1217.
[30] Bawarski, WE; Chidlowsky, E; Bharali, DJ; Mousa, SA. Emerging, Nanopharmaceuticals. *Nanomedicine: Nanotechnology, Biology and Medicine,* 2008, 4, 273-282.
[31] Wei, CM; Liu, NH; Xu, PY; Heller, M; Tomalia, DA; Haynie, DI; Chang, EH; Wang, K; Lee, Y-S; Lyubchenko, YL; et al: From Bench to Bedside: Successful Translational Nanomedicine. *Nanomedicine: Nanotechnology, Biology and Medicine,* 2007, 3, 322-331.
[32] http://nihroadmap.nih.gov/clinicalresearch/overview-translational.asp.
[33] The National Nanotechnology Initiative. Second Assessment and Recommendations of the National Nanotechnology Advisory Panel, April 2008:www.nano.gov /PCAST _NNAP_NNI_Assessment_2008.pdf.
[34] Caruthers, SD; Wickline, SA; Lanza, GM; Nanotechnological Applications in Medicine. *Current Opinion in Biotechnology,* 2007, 18, 26-30.
[35] Moghimi, SM; Hunter, AC; Murray, JC; Nanomedicine: Current Status and Future Prospects. *The FASEB Journal,* 2006, 19, 311-330.
[36] Sahoo, SK; Parveen, S; Panda, JJ. The Present and Future of Nanotechnology in Human Health Care. *Nanomedicine: Nanotechnology, Biology and Medicine,* 2007, 3, 20-31.
[37] American Cancer Society. *Cancer Facts and Figures*, 2007. 2007.
[38] Kim, KY. Nanotechnology Platforms and Physiological Challenges for Cancer Therapeutics.*Nanomedicine: Nanotechnology, Biology and Medicine,* 2007, 3, 103-110.

[39] Koo, YEL; Reddy, GR; Bhojani, M; Schneider, R; Philbert, MA; Rehemtulla, A; Ross, BD; Kopelman, R. Brain Cancer Diagnosis and Therapy with Nanoplatforms.*Adv Drug Delivery Reviews,* 2006, 58, 1556-1577.

[40] Chin CD; Linder V; Sia SK. Lab-on-a-chip Devices for Global Health: Past Studies and Future Opportunities. *Lab on a Chip* 2007, 7, 41-57.

[41] Christodoulides, N; Floriano, PN; Miller, CS; Ebersole, JL; Mohanty, S. et al: Lab-on-a-chip Methods for Point-of-care Measurements of Salivary Biomarkers of Periodontitis.*Ann NY Acad Sci,* 2007, 1098, 411-428.

[42] Hild, WA; Breunig, M; Goepferich, A. Quantum Dots - Nano-sized Probes for the Exploration of Cellular and Intracellular Targeting.*European Journal of Pharmaceutics and Biopharmaceutics,* 2008, 68, 153-168.

[43] Fuller, JE; Zugates, GT; Ferreira, LS; Ow, HS; Nguyen, NN; Wiesner, UB; Langer, RS. Intracellular Delivery of Core - Shell Fluorescent Silica Nanoparticles.*Biomaterials,* 2008, 29, 1526-1532.

[44] El-Sayed, I; Huang, X; Macheret, F; Humstoe, JO; Kramer R; El-Sayed M. Effect of Plasmonic Gold Nanoparticles on Benign and Maglignant Cellular Autofluorescence: A Novel Probe for Fluorescence Based Detection of Cancer. *Technology in Cancer Research and Treatment,* 2007, 6, 403-412.

[45] Sharma, P; Brown, S; Walter, G; Santra, S; Moudgil B. Nanoparticles for Bioimaging.*Adv in Colloid and Interface Science,* 2006, 123, 471-485.

[46] He, J; VanBrocklin, HF; Franc, BL; Seo, Y; Jones, EF. Nanoprobes for Medical Diagnostics: Current Status of Nanotechnology in Molecular Imaging.*Current Nanoscience,* 2008, 4, 17-29.

[47] Burns, A; Ow, H; Weisner, U. Fluorescent Core-shell Silica Nanoparticles: Towards "Lab on a Particle" Architectures for Nanobiotechnology.*Chem Soc Rev,* 2006, 35, 1028-1042.

[48] Tomalia, DA; Reyna, LA; Svenson, S. Dendrimers as Multi-purpose Nanodevices for Oncology Drug Delivery and Diagnostic Imaging.*Biochemical Society Transactions,* 2007, 35, 61-67.

[49] Briones, E; Colino, CI; Lanao, JM. Delivery Systems to Increase the Selectivity of Antibiotics in Phagocytic Cells.*Journal of Controlled Release,* 2008, 125, 210-227.

[50] Hwang, SH; Rait, A; Pirollo, KF; Zhou, O; Yenugonda, VM; Chinigo, GM; Brown, ML; Chang, EH. Tumor-targeting Nanodelivery Enhances the Anticancer Activity of a Novel Quinazolinone Analogue.*Mol Cancer Therapy,* 2008, 7, 559-568.

[51] Sutton, D; Nasongkla, N; Blanco, E; Gao, JM. Functionalized Micellar Systems for Cancer Targeted Drug Delivery.*Pharmaceutical Research,* 2007, 24, 1029-1046.

[52] Conti, M; Tazzari, V; Baccini, C; Pertici, G; Serino, LP; De Giorgi, U. Anticancer Drug Delivery with Nanoparticles.*In Vivo* 2006, 20, 697-701.

[53] Maier-Hauff, K; Rothe, R; Scholz, R; Gneveckow, U; Wust, P; Thiesen, B; Feussner, A; von Deimling, A; Waldoefner, N; Felix, R; Jordan, A. Intracranial Thermotherapy Using Magnetic Nanoparticles Combined with External Beam Radiotherapy: Results of a Feasibility Study on Patients with Glioblastoma Multiforme. *J of Neuro-oncology,* 2007, 81, 53-60.

[54] Economic Costs of Diabetes in the U.S. in 2007.*Diabetes Care,* 2008, 31, 596-315.

[55] American Diabetes Association European Comission: Nanomedicine: Nanotechnology for Health. *European Technology Platform,* 2006.

[56] Cavalcanti, A; Shirinzadeh, B; Kretly, LC. Medical Nanorobotics for Diabetes Control. *Nanomedicine: Nanotechnology, Biology and Medicine,* 2008, 4, 127-138.

[57] Desai, TA; West, T; Cohen, M; Boiarski, T; Rampersaud, A. Nanoporous Microsystems for Islet Cell Replacement.*Adv Drug Delivery Reviews,* 2004, 56, 1661-1673.

[58] Berkowitz, M; O'Leary, PK; Kruse, DL; Harvey, C. *Spinal Cord Injury: an Analysis of Medical and Social Costs,* 1998: ISBN 1-888799-17-X. Demos Medical Publishing.

[59] Harvey, C; Wilson, SE; Greene, CG; Berkowitz, M; Stripling, TE; New Estimates of the Direct Costs of Traumatic Spinal-Cord Injuries - Results of a Nationwide Survey. *Paraplegia,* 1992, 30, 834-850.

[60] Silva GA. Neuroscience Nanotechnology: Progress, Opportunities and Challenges. *Nat Rev Neurosci,* 2006, 7, 65-74.

[61] Tysseling-Mattiace, VM; Sahni, V; Niece, KL; Birch, D; Czeisler, C; Fehlings, MG; Stupp, SI; Kessler, JA. Self-Assembling Nanofibers Inhibit Glial Scar Formation and Promote Axon Elongation after Spinal Cord Injury.*Journal of Neuroscience,* 2008, 28, 3817-3823.

[62] Liu-Snyder, P; Webster, TJ. Developing a New Generation of Bone Cements with Nanotechnology.*Current Nanoscience,* 2008, 4, 111-118.

[63] Khang, D; Lu, J; Yao, C; Haberstroh, KM; Webster, TJ. The Role of Nanometer and Sub-micron Surface Features on Vascular and Bone Cell Adhesion on Titanium. *Biomaterials,* 2008, 29, 970-983.

[64] Chan, CK; Kumar, TSS; Liao, S; Murugan, R; Ngiam, M; Ramakrishman, S. Biomimetic Nanocomposites for Bone Graft Application. *Nanomedicine,* 2006, 1:177-188.

[65] White, AA; Best, SM; Kinloch, IA. Hydroxyapatite-Carbon Nanotube Composites for Biomedical Applications: a Review.*International Journal of Applied Ceramic Technology,* 2007, 4, 1-13.

[66] New Natural Plastic Extends Life of Bone Implant. htp://www.sciencedaily .com/releases/2007/10/071030132322.htm.

[67] Nanomedicine: Nanotechnology for Health, European Technology Platform, Strategic Research Agenda for Nanomedicine, November 2006: http://cortis.europa.eu/ nanotechnology/nanomedicine.htm.

[68] Qian, XM; Peng, XH; Ansari, DO; Yin-Goen, Q; Chen, GZ; Shin, DM; Yang, L; Young, AN; Wang, MD; Nie, SM. In vivo Tumor Targeting and Spectroscopic Detection with Surface-enhanced Raman Nanoparticle Tags. *Nature Biotechnology,* 2008, 26, 83-90.

[69] Leary, SP; Liu, CY; Apuzzo, MLJ. Toward the Emergence of Nanoneurosurgery: Part II Nanomedicine: Diagnostics and Imaging at the Nanoscale Level. *Neurosurgery,* 2006, 58, 805-822.

[70] Huang, XH; El-Sayed, IH; Qian, W; El-Sayed, MA. Cancer Cells Assemble and Align Gold Nanorods Conjugated to Antibodies to Produce Highly Enhanced, Sharp, and Polarized Surface Raman Spectra: A Potential Cancer Diagnostic Marker.*Nano Letters,* 2007, 7, 1591-1597.

[71] Zhang, P; Guo, Y. Surface-Enhanced Raman Scattering Inside Metal Nanoshells. *J Am Chem Soc, ASAP* 2009.

[72] Stewart, ME; Anderton, CR; Thompson, LB; Maria, J; Gray, SK; Rogers, JA; Nuzzo, RG. Nanostructured Plasmonic Sensors.*Chem Rev,* 2008, 108:494-521.
[73] Tang, HW; Yang, XB; Kirkham, J; Smith, DA. Probing Intrinsic and Extrinsic Components in Single Osteosarcoma Cells by Near-infrared Surface-enhanced Raman Scattering.*Anal Chem,* 2007, 79, 3646-3653.
[74] McAlpine, MC; Ahmad, H; Wang, DW; Health, JR. Highly Ordered Nanowire Arrays on Plastic Substrates for Ultrasensitive Flexible Chemical Sensors.*Nature Materials,* 2007, 6, 379-384.
[75] Kim, DH; Ahn, JH; Choi, WM; Kim, HS; Kim, TH; Song, JZ; Huang, YGY; Liu, ZJ; Lu, C; Rogers, JA. Stretchable and Foldable Silicon Integrated Circuits. *Science,* 2008, 320, 507-511.
[76] http://www.alcoholmonitoring.com/.
[77] http://www.echotx.com/.
[78] http://www.flexmedsys.com/index.htm.
[79] Ren, S; Yoon, HR; Kim, SY. Trends in Three-Dimensional Biochips.*Biochip Journal* 2008, 2, 155-159.
[80] Abraham, AM; Kannangai, R; Sridharan, G. Nanotechnology: A new Frontier in Virus Detection in Clinical Practice. *Indian Journal of Me dical Microbiology,* 2008, 26, 297-301.
[81] Bake, KD; Walt, DR. Multiplexed Spectroscopic Detections. *Annual Review of Analytical Chemistry,* 2008, 1, 515-547.
[82] Collard, D; Takeuchi, S; Fujita, H. MEMS Technology for Nanobio Research. *Drug Discovery Today,* 2008, 13, 989-996.
[83] Wimberger-Friedl, R; Nellissen, T; Weekamp; W; van Delft, J; Ansems, W; Prins, M; Megens, M; Dittmer, W; de Witz, C; van Iersel, B. Packaging of Silicon Sensors for Microfluidic Bio-Analytical Applications. *Journal of Micromechanics and Microengineering,* 2009, 19, 015015.
[84] Zengerle, R; Koltay, P; Ducree, J; Microfluidics. An Enabling Technology for the Life Sciences. ISBN 0-7803-8607-8. In *Proceedings of the 2004 International Symposium on Mico-Nanomechatronics and Human Science, 2004 and The Fourth Symposium Micro-Nanomechatronics for Information-Based Society; IMTEK, Freiburg Univ., Freiberg, Germany.* 2004: 1-6.
[85] Sathuluri, RR; Yamamura, S; Tamiya, E. Microsystems Technology and Biosensing. *Advances in Biochemical Engineering/Biotechnology,* 2008, 109, 285-350.
[86] Samel, B; Nock, V; Russom, A; Griss, P; Stemme, G. A Disposable Lab-on-a-chip Platform with Embedded Fluid Actuators for Active Nanoliter Liquid Handling. *Biomedical Microdevices,* 2007, 9, 61-67.
[87] Kakuta, M; Takahashi, H; Kazuno, S; Murayama, K; Ueno, T; Tokeshi, M; Development of the Microchip-based Repeatable Immunoassay System for Clinical Diagnosis. *Measurement Science and Technology,* 2006, 17, 3189-3194.
[88] Virdi, GS; Chutani, RK; Rao, PK; Kumar, S. Fabrication of Low Cost Integrated Micro-capillary Electrophoresis Analytical Chip for Chemical Analysis.*Sensors and Actuators,* 2008, B128, 422-426.
[89] Hunt, HC; Wilkinson, JS. Optofluidic Integration for Microanalysis.*Microfluid Nanofluid,* 2008, 4, 53-79.

[90] Moran-Mirabal, JM; Torres, AJ; Samiee, KT; Baird, BA; Craighead, HG. Cell Investigation of Nanostructures: Zero-mode Waveguides for Plasma Membrane Studies with Single Molecule Resolution. *Nanotechnology,* 2007, 18, 195101.

[91] Weibel, DB; DiLuzio, WR; Whitesides, GM. Microfabrication Meets Microbiology. *Nature Reviews: Microbiology,* 2007, 5, 209-218.

[92] Stern, E; Klemic, JF; Routenberg, DA; Wyrembak, PN; Turner-Evans, DB. Hamilton, AD; LaVan, DA; Fahmy, TM; Reed, MA. Label-free Immunodetection with CMOS-compatible Semiconducting Nanowires.*Nature,* 2007, 445, 519-522.

[93] Kalorama Information: Analytical Chip Technology: U.S. Markets for Lab on a Chip, DNA/Gene, Protein and other Microarrays. *Kalorama publication ID KLI 1393085* 2007.

[94] Christodoulides, N; Floriano, PN; Acosta, SA; Ballard, KLM; Weigum, SE; Mohanty, S; Dharshan, P; Romanovicz, D; McDevitt, JT. Toward the Development of a Lab-on-a-Chip Dual-function Leukocyte and C-Reactive Protein Analysis Method for the Assessment of Inflamation and Cardiac Risk. *Clinical Chemistry,* 2005, 51, 2391-2395.

[95] Reigger, L; Grumann, M; Steigert, J; Lutz, S; Steinert, CP; Mueller, C; Viertel, J; Prucker, O; Ruhe, J; Zengerle, R; Ducree, J. Single-step Centrifugal Hematocrit Determination on a 10-$ Processing Device. *Biomed Microdevices,* 2007, 9, 795-799.

[96] Snider, RM; Ciobanu, M; Rue, AE; Cliffel, DE. A Multiwalled Carbon Nanotube/Dihydropyran Composite Film Electrode for Insulin Detection in a Microphysiometer Chamber. *Analytica Chimica Acta,* 2008, 609, 44-52.

[97] Price, CP; Kricka, LJ. Improving Healthcare Accessibility through Point-of-Care Technologies. *Clinical Chemistry,* 2007, 53, 1665-1675.

[98] Cross, SE; Jin, YS; Rao, J; Gimzewski, JK. Nanomechanical Analysis of Cells from Cancer Patients. *Nature Nanotechnology,* 2007, 2, 780-783.

[99] The Biomonitoring Futures Project. Final Report and Recommendations, Institute for Alternative Futures,2006 : www.altfutures.com/BFP..

[100] Martinez, AW; Phillips, ST; Carrilho, E; Thomas, SW; III, Sindi, H; Whitesides, GM. Simple Telemedicine for Developing Regions: Camera Phones and Paper-Based Microfluidic Devices for Real-Time, Off-Site Diagnosis. *Anal Chem,* 2008, 80, 3699-3707.

[101] Haeberle, S; Zengerle, R. Microfluidic Platforms for Lab-on-a-Chip Applications. *Lab on a Chip,* 2007, 7, 1094-1110.

[102] Kang, LF; Chung, BG; Langer, R; Khademhosseini, A. Microfluidics for Drug Discovery and Development From Target Selection to Product Lifecycle Management. *Drug Discovery Today,* 2008, 13, 1-13.

[103] Gross, S; Piwnica-Worms, D. Molecular Imaging Strategies for Drug Discovery and Development. *Current Opinion on Chemical Biology,* 2006, 10, 334-342.

[104] Ferrari M. Beyond Drug Delivery. *Nature Nanotechnology,* 2008, 3, 131-132.

[105] Hanley, C; Layne, J; Punnoose, A; Reddy, KM,;Coombs, I; Coombs, A; Feris, K; Wingett, D. Preferential Killing of Cancer Cells and Activated Human T Cells Using ZnO Nanoparticles. *Nanotechnology,* 2008, 19, 1-10.

[106] Maguire, AM; Simonelli, F; Pierce, EA; Pugh, EN; Mingozzi, F; Bennicelli, J; Banfi, S; Marshall, KA; Testa, F; Surace, EM. et al: Safety and Efficacy of Gene Transfer for Leber's Congenital Amaurosis. *The New England Journal of Medicine,* 2008, 358:2240-2248.

[107] Byrappa, K; Ohara, S; Adschiri, T. Nanoparticles Synthesis Using Supercritical Fluid Technology - Towards Biomedical Applications. *Advanced Drug Delivery Reviews,* 2008, 60, 299-327.

[108] Hartman, KB; Wilson, LJ; Rosenblum, MG. Detecting and Treating Cancer with Nanotechnology. *Molecular Diagnosis & Therapy,* 2008, 12, 1-14.

[109] Heath, JR; Davis, ME. Nanotechnology and Cancer. *Ann Rev Med,* 2008, 59:251-265.

[110] Leary, SP; Liu, CY; Apuzzo, MLI. Toward the Emergence of Nanoneurosurgery: Part III - Nanomedicine: Targeted Nanotherapy, Nanosurgery, and Progress toward the Realization of Nanoneurosurgery. *Neurosurgery,* 2006, 58, 1009-1026.

[111] Linkov, I; Satterstrom, FK; Corey, LM. Nanotoxicology and Nanomedicine: Making Hard Decisions.*Nanomedicine: Nanotechnology, Biology and Medicine* 2008, 4, 167-171.

[112] Ashammakhi, N; Wimpenny, I; Nikkola, L; Yang, Y. Electrospinning: Methods and Development of Biodegradable Nanofibres for Drug Release. *J Biomedical Nanotechnology,* 2009, 5, 1-19.

[113] Gu, F; Zhang, L; Teply, BA; Mann, N; Wang, A; Radovic-Moreno, AF; Langer, R. Farokhzad OC: Precise Engineering of Targeted Nanoparticles by using Self-assembled Biointegrated Block Copolymers. *PNAS,* 2008, 105, 2586-2591.

[114] Winter, PM; Cai, K; Caruthers, SD; Wickline, SA; Lanza, GM. Emerging Nanomedicine Opportunites with Perfluorocarbon Nanoparticles. *Expert Rev Med Devices,* 2007, 4, 137-145.

[115] Lesniak, WG; Kariapper, MST; Nair, BM; Tan, W; Hutson, A; Balogh, LP; Khan, MK. Synthesis and Characterization of PAMAM Dendrimer-based Multifunctional Nanodevices for Targeting alpha(v)beta(3) Integrins. *Bioconjgate Chem,* 2007, 18, 1148-1154.

[116] Lowery, AR; Gobin, AM; Day, ES; Halas, NJ; West, JL. Immunonanoshells for Targeted Photothermal Ablation of Tumor Cells. *International Journal of Nanomedicine,* 2006, 1, 149-154.

[117] Fu, K; Sun, J; Bickford, LR; Lin, AWH; Halas, NJ; Yu, TK; Drezek, RA. Measurement of Immunotargeted Plasmonic Nanoparticles' Cellular Binding: A Key Factor in Optimizing Diagnostic Efficacy. *Nanotechnology,* 2008, 19, 045103.

[118] Rapoport, N; Gao, ZG; Kennedy, A. Multifunctional Nanoparticles for Combining Ultrasonic Tumor Imaging and Targeted Chemotherapy. *J Natl Cancer Inst,* 2007, 99, 1095-1106.

[119] Piao, Y; Kim, J; Bin Na, H; Kim, D; Baek, JS; Ko, MK; Lee, JH; Shokouhimehr, M; Hyeon, T. Wrap-Bake-Peel Process for Nanostructural Transformations from ß-FeOOH Nanorods to Biocompatible Iron Oxide Nanocapsules. *Nature Materials,* 2008, 7, 242-247.

[120] Myc, A; Majoros, IJ; Thomas, TP; Baker, JR. Dendrimer-based Targeted Delivery of an Apoptotic Sensor in Cancer Cells. *Biomacromolecules,* 2007, 8, 13-18.

[121] Birch, DG; Liang, FQ. Age-related Macular Degeneration: as Target for Nanotechnology Derived Medicines. *International Journal of Nanomedicine,* 2007, 2, 65-77.

[122] Chumakova, OV; Liopo, AV; Andreev, VG; Cicenaite, I; Evers, BM; Chakrabarty, S; Pappas, TC; Esenaliev, RO. Composition of PLGA and PEI/DNA Nanoparticles

Improves Ultrasound-mediated Gene Delivery in Solid Tumors in vivo.*Cancer Letters,* 2008, 261, 215-225.
[123] Gill, S; Lobenberg, R; Ku, T; Azarmi, S; Roa, W; Prenner, EJ. Nanoparticles: Characteristics, Mechanisms of Action, and Toxicity in Pulmonary Drug Delivery - a Review. *J Biomedical Nanotechnology,* 2007, 3, 107-119.
[124] Torchilin, VP. Multifunctional Nanocarriers. *Adv Drug Delivery Reviews,* 2006, 58, 1532-1555.
[125] Jin, S; Ye, KM. Nanoparticle-Mediated Drug Delivery and Gene Therapy. *Biotechnol Prog,* 2007, 23, 32-41.
[126] Cho, YW; Park, SA; Han, TH; Son, DH; Park, JS; Oh, SJ; Moon, DH; Cho, KJ; Ahn, CH; Byun, Y. et al: In vivo Tumor Targeting and Radionuclide Imaging with Self-assembled Nanoparticles: Mechanisms, Key Factors, and their Implications. *Biomaterials,* 2007, 28, 1236-1247.
[127] Balogh, L; Nigavekar, SS; Nair, BM; Lesniak, W; Zhang, C; Sung, LY; Kariapper, MST; El-Jawahri, A; Llanes, M; Bolton, B. et al: Significant Effect of Size on the in-vivo Biodistribution of Gold Composite Nanodevices in Mouse Tumor Models. *Nanomedicine: Nanotechnology, Biology and Medicine,* 2007, 3, 281-296.
[128] Roovers, RC; van Dongen, GAMS. Henegouwen PMPVE: Nanobodies in Therapeutic Applications. *Current Opinion on Molecular Therapeutics,* 2007, 9, 327-335.
[129] Bae, YH. Cancer Targeting Paradigm: Does it Need to be Shifted?*Nanomedicine: Nanotechnology, Biology and Medicine* 2007, 3, 344-344.
[130] Leroueil, PR; Berry, SA; Duthie, K; Han, G; Rotello, VM; McNerny, DQ; Baker, JR; Orr, BG; Holl, MMB. Wide Varieties of Cationic Nanoparticles Induce Defects in Supported Lipid Bilayers. *Nano Letters,* 2008, 8, 420-424.
[131] Huff, TB; Tong, L; Zhao, Y; Hansen, MN; Cheng, JX; Wei, A. Hyperthermic Effects of Gold Nanorods on Tumor Cells. *Nanomedicine,* 2007, 2, 125-132.
[132] Kalele, SA; Tiwari, NR; Gosavi, SW; Kulkarni, SK. Plasmon-assisted Photonics at the Nanoscale. *Journal of Nanophotonics,* 2007, 1, 012501.
[133] Gannon, CJ; Cherukuri, P; Yakobson, BI; Cognet, L; Kanzius, JS; Kittrell, C; Weisman, RB; Pasquali, M; Schmidt, HK; Smalley, RE; Curley, SA. Carbon Nanotube-enhanced Thermal Destruction of Cancer Cells in a Noninvasive Radiofrequency Field. *Cancer,* 2007, 110, 2654-2665.
[134] Christie, JS; Kompella, UB. Ophthalmic Light Sensitive Nanocarrier Systems. *Drug Discovery Today,* 2008, 13, 124-134.
[135] Gao, D; Agayan, RR; Xu, H; Philbert, MA; Kopelman, R. Nanoparticles for Two-photon Photodynamic Therapy in Living Cells. *Nano Letters,* 2006, 6, 2383-2386.
[136] Chen, HT; Kaminski, MD; Liu, XQ; Mertz, CJ; Xie, YM; Torno, MD; Rosengart, AJ. A Novel Human Detoxification System Based on Nanoscale Bioengineering and Magnetic Separation Techniques. *Medical Hypotheses,* 2007, 68, 1071-1079.
[137] Barnes, CP; Sell, SA; Boland, ED; Simpson, DG; Bowlin, GL. Nanofiber Technology: Designing the Next Generation of Tissue Engineering Scaffolds. *Adv Drug Delivery Reviews,* 2007, 59, 1413-1433.
[138] Badylak, SF; Freytes, DO; Gilbert, TW. Extracellular Matrix as a Biological Scaffold Material: Structure and Function. *ACTA Biomaterialia,* 2009, 5, 1-13.
[139] Venugopal, J; Low, S; Choon, AT. Ramakrishna S: Interaction of Cells and Nanofiber Scaffolds in Tissue Engineering. *J Biomedical Materials Research,* 2008, 84B, 34-48.

[140] Venugopal, J; Prabhakaran, MP; Low, S; Choon, AT; Zhang, YZ; Deepika, G. Ramakrishna S: Nanotechnology for Nanomedicine and Delivery of Drugs. *Current Pharmaceutical Design,* 2008, 14, 2184-2200.

[141] Hunley, MT; Long, TE. Electrospinning Functional Nanoscale Fibers: a Perspective for the Future. *Polymer International,* 2008, 57, 385-389.

[142] Powell, HM; Supp, DM; Boyce, ST. Influence of Electrospun Collagen on Wound Contraction of Engineered Skin Substitutes. *Biomaterials,* 2008, 29, 834-843.

[143] Martins, A; Araujo, JV; Reis, RL; Neves, NM. Electrospun Nanostructured Scaffolds for Tissue Engineering Applications. *Nanomedicine,* 2007, 2, 929-942.

[144] Liao, S; Li, BJ; Ma, ZW; Wei, H; Chan, C. Ramakrishna S: Biomimetic Electrospun Nanofibers for Tissue Regeneration. *Biomedical Materials,* 2006, 1, R45-R53.

[145] Ma, ZW; Kotaki, M; Inai, R; Ramakrishna, S. Potential of Nanofiber Matrix as Tissue-engineering Scaffolds. *Tissue Enginering,* 2005, 11, 101-109.

[146] Tsonchev, S; Niece, KL; Schatz, GC; Ratner, MA; Stupp, SI. Phase Diagram for Assembly of Biologically-Active Peptide Amphiphiles. *J Phys Chem,* 2008, B112, 441-447.

[147] Hartgerink, JD; Beniash, E; Stupp, SI. Self-Assembly and Mineralization of Peptide-Amphiphile Nanofibers. *Science,* 2001, 294, 1684-1688.

[148] Chau, Y; Luo, Y; Cheung, ACY; Nagai, Y; Zhang, SG; Kobler, JB; Zeitels, SM; Langer, R. Incorporation of a Matrix Metalloproteinase-sensitive Substrate into Self-assembling Peptides - A Model for Biofunctional Scaffolds. *Biomaterials,* 2008, 29, 1713-1719.

[149] Tao, SL; Desai, TA. Aligned Arrays of Biodegradable Poly(epsilon-caprolactone) Nanowires and Nanofibers by Template Synthesis. *Nano Letters,* 2007, 7, 1463-1468.

[150] Erisken, C; Kalyon, DM; Wang, HJ. A Hybrid Twin Screw Extrusion/Electrospinning Method to Process Nanoparticle-incorporated Electrospun Nanofibres. *Nanotechnology,* 2008, 19, 165302.

[151] Hosseinkhani, H; Hosseinkhani, M; Khademhoseini, A. Tissue Regeneration Through Self-assembled Peptide Amphiphile Nanofibers. *Yakhteh Medical Journal,* 2006, 8, 204-209.

[152] Capito, RM; Azevedo, HS; Velichko, YS; Mata, A; Stupp, SI. Self-assembly of Large and Small Molecules into Hierarchically Ordered Sacs and Membranes. *Science,* 2008, 319, 1812-1816.

[153] Wang, XM; Horii, A; Zhang, SG. Designer Functionalized Self-assembling Peptide Nanofiber Scaffolds for Growth, Migration, and Tubulogenesis of Human Umbilical Vein Endothelial Cells. *Soft Matter,* 2008, 4, 2388-2395.

[154] Audette, GF; Hazes, B. Development of Protein Nanotubes from a Multi-purpose Biological Structure. *J Nanosci Nanotechnol,* 2007, 7, 2222-2229.

[155] Sargeant, TD; Guler, MO; Oppenheimer, SM; Mata, A; Satcher, RL; Dunand, DC; Stupp, SI. Hybrid Bone Implants: Self-assembly of Peptide Amphiphile Nanofibers with Porous Titanium. *Biomaterials,* 2008, 29, 161-171.

[156] Khang, D; Park, GE; Webster, TJ. Enhanced Chondrocyte Densities on Carbon Nanotube Composites: the Combined Role of Nanosurface Roughness and Electric Stimulation. *Journal of Biomedical Materials Research Part A,* 2008, 86A, 253-260.

[157] Popat, KC; Leoni, L; Grimes, CA; Desai, TA. Influence of Engineered Titania Nanotubular Surfaces on Bone Cells. *Biomaterials,* 2007, 28, 3188-3197.

[158] Venugopal, J; Vadgama, P; Kumar, TSS; Ramakrishna, S. Biocomposite Nanofibres and Osteoblasts for Bone Tissue Engineering. *Nanotechnology,* 2007, 18, 055101.

[159] Ballard, JD; Dulgar-Tulloch, AJ; Siegel, RW. Nanophase Materials. In *Encyclopedia of Biomedical Engineering.Volume* 4. p.2489-2507, Wiley, ISBN-10 142007802X, 2006.

[160] http://solutions.3m.com/wps/portal/3M/en_US/3M-ESPE/dental-professionals/products/category/direct-restorative/filtek-supreme-plus/.

[161] Randall, CL; Leong, TG; Bassik, N; Gracias, DH. 3D Lithographically Fabricated Nanoliter Containers for Drug Delivery. *Adv Drug Delivery Reviews,* 2007, 59, 1547-1561.

[162] Chen, B; Wei, J; Tay FEH, Wong, YT; Iliescu, C. Silicon Microneedle Array with Biodegradable Tips for Transdermal Drug Delivery. *Microsystem Technologies-Micro-and Nanosystems-Information Storage and Processing Systems,* 2008, 14, 1015-1019.

[163] Nisar, A; Aftulpurkar, N; Mahaisavariya, B; Tuantranont, A. MEMS-based Micropumps in Drug Delivery and Biomedical Applications. *Sensors and Actuators,* 2008, B130, 917-942.

[164] Heller, A; Potentially Implantable Miniature Batteries. *Analytical and Bioanalytical Chemistry,* 2006, 385, 469-473.

[165] Li, PF, Bashirullah, R. A Wireless Power Interface for Rechargeable Battery Operated Medical Implants. *IEEE Transactions on Circuits and Systems II - Express Briefs,* 2007, 54, 912-916.

[166] Riistama, J; Vaisanen, J; Heinisuo, S; Harjunpaa, H; Arra, S; Kokko, K; Maentylae, M; Kaihilahti, J; Heino, P; Kellomaki, M. et al: Wireless and Inductively Powered Implant for Measuring Electrocardiogram. *Medical and Biological Engineering and Computing,* 2007, 45, 1163-1174.

[167] Wang, XD; Song, JH; Liu, J; Wang, ZL. Direct-current Nanogenerator Driven by Ultrasonic Waves. *Science,* 2007, 316, 102-105.

[168] Qin, Y; Wang, XD; Wang, ZL. Microfibre-nanowire Hybrid Structure for Energy Scavenging. *Nature,* 2008, 451, 809-813.

[169] Laocharoensuk, R; Burdick, J; Wang, J. Carbon-nanotube-induced Acceleration of Catalytic Nanomotors. *ACS Nano,* 2008, 2, 1069-1075.

[170] Wang, ZL. The New Field of Nanopiezotronics. *Materials Today,* 2007, 10, 20-28.

[171] Grill, WM. NAKFI Smart Prosthetics: Exploring Assistive Devices for the Body and Mind. *Expert Rev Med Devices,* 2007, 4, 107-108.

[172] National Academies Keck Futures Initiative: Smart Prosthetics: Exploring Assistive Devices for the Body and Mind: Task Group Summaries. *The National Academies Press, ISBN-10 0-309-10466-1*, 2007.

[173] Lebedev MA, Nicolelis MAL: Brain-machine Interfaces: Past, Present, Future. *Trends Neurosci,* 2006, 29, 536-546.

[174] Elder, JB; Liu, CY; Apuzzo, MLJ. Neurosurgery in the Realm of 10(-9), Part 2: Applications of Nanotechnology to Neurosurgery - Present and Future. *Neurosurgery,* 2008, 62, 269-284.

[175] Silva, GA. Nanotechnology Approaches for Drug and Small Molecule Delivery Across the Blood Brain Barrier. *Surgical Neurology,* 2007, 67, 113-116.

[176] Wise, KD; Bhatti, PT; Wang, JB; Friedrich, CR. High-density Cochlear Implants with Position Sensing and Control. *Hearing Research,* 2008, 242, 22-30.

[177] Duke, T. The Power of Hearing. *Physics World,* 2002, 15, 29-33.

[178] Zaghloul, KA; Baohen, K. A Silicon Retina that Reproduces Signals in the Optic Nerve. *J Neural Eng,* 2006, 3, 257-267.
[179] Cohen, ED. Prosthetic Interfaces with the Visual System: Biological Issues. *J Neural Eng,* 2007, 4, R14-R31.
[180] Pappas, TC; Wickramanyake, WMS; Jan, E; Motamedi, M; Brodwick, M; Kotov, NA. Nanoscale Engineering of a Cellular Interface with Semiconductor Nanoparticle Films for Photoelectric Stimulation of Neurons. *Nano Letters,* 2007, 7, 513-519.
[181] Kuiken, TA; Marasco, PD; Lock, BA; Harden, RN; Dewald, JPA. Redirection of Cutaneous Sensation from the Hand to the Chest Skin of Human Amputees with Targeted Reinnervation. *PNAS,* 2007, 104, 20061-20066.
[182] Sharp, AA; Panchawagh, HV; Ortega, A; Artale, R; Richardson-Burns, S. Finch, DS; Gall, K; Mahajan, RL; Restrepo, D. Toward a Self-Deploying Shape Memory Polymer Neuronal Electrode. *J Neural Eng,* 2006, 3, L23-L30.
[183] Cabanlit, M; Maitland, D; Wilson, T; Simon, S; Wun, T; Gershwin, ME. Van de Water J: Polyurethane Shape-memory Polymers Demonstrate Functional Biocompatibility in vitro. *Macromolecular Bioscience,* 2007, 7, 48-55.
[184] Mazzatenta, A; Giugliano, M; Campidelli, S; Gambazzi, L; Businaro, L; Markram, H; Prato, M; Ballerini, L. Interfacing Neurons with Carbon Nanotubes: Electrical Signal Transfer and Synaptic Stimulation in Cultured Brain Circuits. *Journal of Neuroscience,* 2007, 27, 6931-6936.
[185] Ouyang, M; Huang, JL; Lieber, CM. Fundamental Electronic Properties and Applications of Single-walled Carbon Nanotubes.*Accounts of Chemical Research,* 2002, 35, 1018-1025.
[186] Moxon, KA; Hallman, S; Aslani, A; Kalkhoran, NM; Lelkes, PI. Bioactive Properties of Nanostructured Porous Silicon for Enhancing Electrode to Neuron Interfaces. *J Biomater Sci Polmer,* 2007, 18, 1263-1281.
[187] Zhao, Y; Larimer, P; Pressler, RT; Strowbridge, BW; Burda, C. Wireless Activation of Neurons in Brain Slices Using Nanostructured Semiconductor Photoelectrodes. *Angewandte Chemie International Edition,* 2009, 48, 2407-2410.
[188] Schanze, T; Hesse, L; Lau, C; Greve,N; Haberer, W; Kammer, S; Doerge, T. Rentzos A, Stieglitz T: An Optically Powered Single-channel Stimulation Implant as Test System for Chronic Biocompatibility and Biostability of Miniaturized Retinal Vision Prostheses. *IEEE Trans of Biomedical Engn,* 2007, 54, 983-992.
[189] Mogilner, A; Wollman, R; Marshall, WF. Quantitative Modeling in Cell Biology: What is it Good for? *Developmental Cell,* 2006, 11, 279-287.
[190] Bruggeman, FJ; Weserhoff, HV. Approaches to Biosimulation of Cellular Processes. *J Biological Physics,* 2006, 32, 273-288.
[191] Assmus, HE; Herwig, R; Cho, KH; Wolkenhauer, O. Dynamics of Biological Systems: Role of Systems Biology in Medical Research. *Expert Rev of Molecular Diagnostics,* 2006, 6, 891-902.
[192] Zlatanova, J; van Holde, K. Single Molecule Biology: What is it and how Does it Work?*Molecular Cell,* 2006, 24, 317-329.
[193] California Institute of Technology: Institute for Systems Biology; http://www. systemsbiology.org/Intro to ISB and Systems Biology/Systems Biology the 21st Century Science.

[194] van den Heuvel, MGL; Dekker, C. Motor Proteins at Work for Nanotechnology. *Science,* 2007, 317, 333-336.
[195] Strychalski, EA; Stavis, SM; Craighead, HG. Non-planar Nanofluidic Devices for Single Molecule Analysis Fabricated Using Nanoglassblowing. *Nanotechnology,* 2008, 19, 315301.
[196] Rhee, M; Burns, M. Nanopore Sequencing Technology: Research Trends and Applications. *Trends in Biotechnology,* 2006, 24, 580-586.
[197] Rhee, M; Burns, MA. Nanopore Sequencing Technology: Nanopore Preparations. *Trends in Biotechnology,* 2007, 25, 174-181.
[198] Mannion, JT; Craighead, HG. Nanofluidic Structures for Single Biomolecule Fluorescent Detection.*Biopolymers,* 2007, 85, 131-143.
[199] Clark, J; Wu, H-C; Jayasinghe, L; Patel, A; Reid, S; Bayley, H. Continuous Base Identification for Single-Molecule Nanopore DNA Sequencing.*Nature Nanotechnology Advanced Online Publication (22 Feb 2009)* 2009.
[200] Stanford Microfluidics Foundry: http://thebigone.stanford.edu/foundry.
[201] Wikswo, JP; Prokop, A; Baudenbacher, F; Cliffel, D; Csukas, B; Velkovsky, M. Engineering Challenges of BioNEMS: the Integration of Microfluidics, Micro- and Nanodevices, Models and External Control for Systems Biology. *IEEE Proc Nanobiotechnology,* 2006, 153, 81-101.
[202] Weibel, DB; Whitesides, GM. Applications of Microfluidics in Chemical Biology. *Current Opinion on Chemical Biology,* 2006, 10, 584-591.
[203] Zhang, XL; Yin, HB; Cooper, JM; Haswell, SJ. A Microfluidic-based System for Analysis of Single Cells Based on Ca2+ Flux. *Electrophoresis,* 2006, 27, 5093-5100.
[204] Marcus, JS; Anderson, WF; Quake, SR. Microfluidic Single-cell mRNA Isolation and Analysis. *Analytical Chemistry,* 2006, 78, 3084-3089.
[205] Schrlau, MG; Falls, EM; Ziober, BL; Bau, HH. Carbon Nanopipettes for Cell Probes and Intracellular Injection. *Nanotechnology,* 2008, 19, 1-4.
[206] Schmidt, R; Wurm, CA; Jakobs, S; Engelhardt, J; Egner, A; Hell, SW. Spherical Nanosized Focal Spot Unravels the Interior of Cells. *Nature Methods,* 2008, 5, 539-544.
[207] Mannion, JT; Craighead, HG. Nanofluidic Structures for Single Biomolecule Fluorescent Detection. *Biopolmers,* 2006, 85, 131-143.
[208] Mannix, RJ; Kumar, S; Cassiola, F; Montoya-Zavala, M; Feinstein, E; Prentiss, M, Ingber, DE. Nanomagnetic Actuation of Receptor-mediated Signal Transduction. *Nature Nanotechnology,* 2008, 3, 36-40.
[209] Reed, J; Frank, M; Troke, JJ; Schmit, J; Han, S; Teitell, MA; Gimzewski, JK. High Throughput Cell Nanomechanics with Mechanical Imaging Interferometry. *Nanotechnology,* 2008, 19, 235101.
[210] Addae-Mensah, KA; Wikswo, JP. Measurement Techniques for Cellular Biomechanics In Vitro. *Exp Biol Med,* 2008, 233, 792-809.
[211] Schrlau, MG; Brailoiu, E; Patel, S; Gogotsi, Y; Dun, NJ; Bau, HH. Carbon Nanopipettes Characterize Calcium Release Pathways in Breast Cancer Cells. *Nanotechnology,* 2008, 19, 325102.
[212] Bystrenova, E; Jelitai, M; Tonazzini, I; Lazar, AN; Huth, M; Stoliar, P; Dionigi, C; Cacace, MG; Nickel, B; Madarasz, E; Biscarini, F. Neural Networks Grown on Organic Semiconductors. *Advanced Functional Materials,* 2008, 18, 1751-1756.

[213] Patolsky, F; Timko, BP; Yu, GH; Fang, Y; Greytak, AB; Zheng, GF; Lieber, CM. Detection, Stimulation and Inhibition of Neuronal Signals with High-density Nanowire Transistor Arrays. *Science,* 2006, 313, 1100-1104.
[214] http://nano.caltech.edu/research_index.htm.
[215] Neeves, KB; Diamond, SL. A Membrane-based Microfluidic Device for Controlling the Flux of Platelet Agonists into Flowing Blood. *Lab on a Chip,* 2008, 8, 701-709.
[216] Hansma, PK; Turner, PJ; Fantner, GE. Bone Diagnostic Instrument. *Review of Scientific Instruments,* 2006, 77, 075105.
[217] Hansma, P; Turner, P; Drake, B; Yurtsev, E; Proctor, A; Matthews, P; Lelujian, J; Randall, C; Adams, J; Jungmann, R. et al: The Bone Diagnostic Instrument II: Indentation Distance Increase. *Review of Scientific Instruments,* 2008, 79, 064303.
[218] Freedman, JR; Mattia, D; Korneva, G; Gogotsi, Y; Friedman, G. Fontecchio, AK. Magnetically Assembled Carbon Nanotube Tipped Pipettes. *Appl Phys Lett* 2007, 90:103108.
[219] Haga, Y; Matsunaga, T; Makishi, W; Totsu, K; Mineta, T; Esashi, M. Minimally Invasive Diagnostics and Treatment Using Micro-nano Machining. *Minimally Invasive Therapy & Allied Technologies,* 2006, 15, 218-225.
[220] Roe, D; Karandikar, B,;Bonn-Savage, N; Gibbins, B; Roullet, JB. Antimicrobial Surface Functionalization of Plastic Catheters by Silver Nanoparticles. *Journal of Antimicrobial Chemotherapy,* 2008, 61, 869-876.
[221] Cao, Q; Kim, HS; Pimparkar, N; Kulkarni, JP; Wang, CJ; Shim, M; Roy, K; Alam, MA & Rogers, JA. Medium-scale Carbon Nanotube Thin-film Integrated Circuits on Flexible Plastic Substrates. *Nature,* 2008, 454, 495-500.
[222] Kim, DH; Rogers, JA. Stretchable Electronics: Materials Strategies and Devices. *Advanced Materials,* 2008, 20, 4887-4892.
[223] Branton, D; Deamer, DW; Marziali, A; Bayley, H; Benner, SA; Butler, T. et al. The Potential and Challenges of Nanopore Sequencing. *Nature Biotechnology,* 2008, 26, 1146-1153.
[224] http://www.nibib.nih.gov/Research/ProgramAreas/MicroNanoPlatforms.
[225] http://www.nibib.nih.gov/Research/ProgramAreas/Nanotech.
[226] http://nano.cancer.
[227] http://www.nhlbi-pen.net/default.php.
[228] http://ntp.niehs.nih.gov/?objectid=7E6B19D0-BDB5-82F8-FAE73011304F542A.
[229] http://www.ncrr.nih.gov/.
[230] http://www.nnin.org/nnin_overview.html.
[231] Sanhai, WR; Sakamoto JH; Canady, R; Ferrari M. Seven Challenges for Nanomedicine. *Nature Nanotechnology,* 2008, 3:242-244.
[232] Hodgins, D; Bertsch, A; Post, N; Frischholz, M; Volckaerts, B; Spensley, J; Wasikiewicz, JM; Higgins, H; von Stetten, F; Kenney, L. Healthy Aims: Developing New Medical Implants and Diagnostic Equipment. *IEEE Pervasive Computing,* 2008, 7, 14-21.

In: Nanoscience, Nanomanufacturing and Nanoelectronics ISBN: 978-1-61761-816-1
Editor: Joseph G. Devin

Chapter 2

CROSS-INDUSTRY ISSUES IN NANOMANUFACTURING

National Institute of Standards and Technology

Keywords: nanomanufacturing, nanomaterials, nanoparticles, surfaces, interfaces, bonding, composites, separations, fractions, instrumentation, characterization, measurement, modeling, simulation, performance properties, consortium

ACKNOWLEDGMENTS

The sponsors wish to thank the principal authors of this chapter and all the participants at the workshop held on May 20-22, 2008 at the National Institute of Standards and Technology in Gaithersburg, Maryland. A complete list is provided in Appendix A. The presentations and discussions at that workshop provided the foundation for this chapter.

Special thanks go to the organizing committee, the breakout session leaders (listed below), and to all those who contributed to the writing of this chapter (noted at the beginning of each section). Thanks to Anne Chaka (NIST), Mike Garner (Intel), Michael Gaitan (NIST), Bob Gelman (Hercules), Ehr Ping Huangfu (DOE), Phil Jones (Imerys), Steve Masia (Sappi), Mike Postek (NIST), Dianne Poster (NIST), Gerard Closset (Agenda 2020 Technology Alliance of AF&PA), and John Cowie (Agenda 2020 Technology Alliance of AF&PA), who played leading roles in organizing the workshop.

Surfaces, Interfaces, and Non-bonded Interactions of Nanomaterials:

- Gary Fletcher (BD)
- Christine Mahoney (NIST)

Nanotechnology-enabled Composites and Matrices:

- Ted Wegner (USDA Forest Service)
- World Nieh (USDA Forest Service)

- Gale Holmes (NIST)

Separations and Fractionation:

- Michael Gaitan (NIST)
- Dierk Roessner (Wyatt Technology Corp)

Thanks to NIST conference services staff for organizing the logistics and making sure the workshop went smoothly.

Thanks to Joan Pellegrino and the staff of Energetics Incorporated of Columbia, Maryland, who assisted with the design, coordination, and facilitation of the workshop, and in developing this workshop report.

We thank the members of the Subcomittee on Nanoscale Science, Engineering, and Technology for their helpful input to the development of this workshop topic. We would additionally like to thank the National Nanotechnology Coordination Office staff for valuable guidance in the formulation of the workshop and report. Input and support from Philip Lippel and Liesl Heeter merit ourthanks. In this same vein, we thank Geoff Holdridge, who was especially helpful in providing advice and assistance.

The final copy editing of this chapter and its productionwas accomplished under considerable pressure by Donna Kimball(NIST), who made it possible for us to meet our publication goal.

EXECUTIVE SUMMARY

An increasingly competitive global market is driving industry to identify more effective ways to meet customer needs by adding value and reducing costs. Nanotechnology offers the promise of dramatically increasing competitiveness by removing long-standing performance limitations. In addition, nanotechnology is expected to revolutionize manufacturing processes and products across many high-value industry sectors, such as medical technology, electronics, aerospace, automotive, chemical, food and dietary supplements, forest products, and pharmaceuticals. Global investment in nanotechnology rose to nearly $13.5 billion in 2007, with a global market predicted by Lux Research to grow from $147 billion in 2007 (see Exhibit 1-a) to over several trillion dollars in the next decade. But significant technical barriers inhibit the discovery, high-volume manufacture and widespread commercialization of nanotechnology-enabled products needed to fuel this growth. The United States Government has invested over $10 billion over the last ten years and is a long-standing leader in nanotechnology funding. Other nations are now rapidly increasing their investments and eroding this lead.

Challenges to the discovery, manufacture, and commercialization of nanotechnology arise because nanoscale structures are at or beyond the limits of our current capabilities to characterize, manipulate, model, and control. The requisite scientific, measurement and data infrastructure is still in its infancy. Hence, addressing the barriers tothe discovery, manufacture, commercialization of nanotechnology-enabled products is beyond the scope of individual companies. While each industry faces its own set of challenges in harnessing the potential benefits of nanotechnology, there are issues currently hindering widespread industrial applications that are shared by all, and the solutions to these problems will have the greatest effect on establishing U.S. leadership in manufacturing nanotechnology-enabled products. For example, both the medical and the aerospace industries have safety and performance concerns, as well as multimillion dollar testing costs per formulation, that dictate nanomaterials be highly purified, uniform, and consistent in large quantities, which is very difficult to achieve.

Hence, this workshop was organized at the National Institute of Standards and Technology (NIST) with the unique purpose of identifying common problems and common opportunities where focused research and collaboration will deliver high-impact solutions to solve fundamental challenges across many industrial sectors. The organizing committee worked for over a year prior to the workshop, holding small workshops and meetings with stakeholders, to identify the top three priorities for crosscutting issues in nanomanufacturing. These three breakout topics are shown in Exhibit E-1. All the topic areas are strongly dependent on a fundamental understanding of the physics and chemistry of surfaces and interfaces. Within each of these topics, common issues included—although were not limited to—measurement, characterization, modeling, performance properties, and environment, health,and safety concerns.

Exhibit E-1. Description of Cross-Industry Nanomanufacturing Workshop Breakout Topics

Surfaces, Interfaces and Non-Bonded Interactions of Nanomaterials	**Nanotechnology-enabled Composites and Matrices**	**Separations and Fractionation**
Identify fundamental science needs and measurement capabilities required to understand and control the surface-dependent properties of nanomaterials such as dispersion, aggregation, and adhesion at their interface with a matrix, with an emphasis on non-covalent bonding interactions, to enable materials by design and overcome challenges in commercial production.	Identify fundamental science needs and measurement capabilities required to understand and control multiple properties of nanocomposites by design to enable superior performance and reduce cost.	Identify fundamental science needs and measurement capabilities required to characterize nanomaterials and enhance their separation and fractionation to address challenges in commercial production of uniform, high quality, stable, and consistent (reproducible) nanomaterials in high volume and at high throughput.

The *Cross-Industry Issues in Nanomanufacturing Workshop* was held on May20-22, 2008 at NIST, Gaithersburg, MD. Numerous thought leaders from industry, government agencies, and universities came to the workshop with applications in mind that could immediately benefit from investment in common solutions at a fundamental, precompetitive level. The following key questions common to all industrial sectorswere posed:

- How can a company determine if nanotechnology offers sufficient competitive advantage to warrant the cost of innovation?

- How can developers predict the value nanotechnology will add to their new product in the face of numerous uncertainties in the manufacturing chain from the properties of the constituent nanomaterials to the performance of the final product?
- How as a nation, can we manage risk to make nanotechnology competitive with other technologies and reap its enormous potential benefit?
- How can we obtain the health and environmental information needed to protect workers and guarantee the safety of end-products?
- What means are available to mitigate the uncertain development cost of breakthrough nanotechnology-enabled solutions?

In addition, the workshop participants emphasized the need for greater investment in fundamental science directed towards industrial problems to enable companies to determine the potential added value of nanotechnology-enabled products and ensure product quality, consistency, and safety. Advances in fundamental science have the advantage of being precompetitive and the most easily leveraged across industrial sectors.

The following (non-prioritized) cross-cutting needs emerged from the workshop:

- The characterization of nanomaterials and their properties in isolation and in matrices, with respect to size, shape, charge, chemistry, uniformity, and distribution. Measurement science needs to be developed that is capable of both nanoscale resolution and statistical significance in high-throughput modes,
- The creation of an accurate database of nanomaterial properties and their nanomanufacturing process conditions to enable materials and processes by design and reduce trial and error,
- The development of validated modeling capability to predict multiple properties of nanomaterials in complex matrices to reduce the time and cost to develop new products and processes,
- The development of nanomaterials with high-value properties that can't be achieved without the novel properties of the nanotechnology components,
- The identification of factors controlling biological interactions of nanoparticles—free and in matrices—to accurately assess and predict environmental, health and safety risks,
- The articulation of reference materials, reference property data, and reference measurement requirements needs,
- The characterization of scale-up challenges for both process control and high-throughput manufacturing and measurement capabilities,
- The challenge to bridge the gap between academic research and commercial needs,
- The development of standards for nanomanufacturing process terminology, nanomaterials specifications, nanomaterial reference standards, and EHS controls.

Workshop participants agreed that the common issues identified in this workshop are only a beginning and that considerable follow-up work is required. These future efforts should be focused on the following:

Exhibit E-2. Workshop Results: Barriers and Priority Cross-Industry Solutions to Advance Nanomanufacturing

Barriers	Priority Solutions
Surface/Interfaces and Non-Bonded Interactions of Nanomaterials	
• **Regulatory/Legal/ Organizational** • **Characterization and Measurement** • **Modeling** • **EH&S** • **Performance Properties**	• Predict and control toxicityof nanoparticles • Develop reference systems relating to market needs • Establish network of labs that test and develop standards for nanomaterial characterization • Research on surface interface properties relevant to performance • Collect list of existing models, measurement tools, and biological interaction assays • Identify or develop surrogate measurement tools for on-line measurements that relate to research tools • Integrate industry needs and academic funding—mechanism and framework to translate commercial needs into fundamental science questions • Correlate fundamental nanoparticle surface understanding to performance priorities • Improved computational models for nanomaterial surface interactions • Measurement techniques and predictive models interaction: identify gaps and devise solution development • Increase collaboration in modeling and experimentation
Nanotechnology-enabled Composites and Matrices	
• **Value** • **Multi-Scale Modeling** • **Characterization** • **Measurement** • **Performance Properties** • **EH&S**	• Nanomaterials informaticsand knowledge base to accelerate progress • Develop methodology to control reactionsand bonding of nanoscale particles within a matrix and across interfaces • Build value proposition for use of nanomanufacturing to meet customer needs • Develop tool kit enabling measurement at nano-to micro-scale (including dispersion of nanoparticles within a matrix and at interfaces) • Predictably scale up lab-scale processes to manufacturing-scale for production of nanomaterials and products • Optimize manufacturing processes • Coordination of research on nanocomposites and matrices across industry, universities, Government, and Federallaboratories
Separations and Fractionation	
• **Measurement Methods** • **Removal of Non-Nanomaterials** • **Manufacturability** • **Purification** • **Characterization**	• Develop parallel microfluidic methods • Build high-throughput production/scalable methods • Enable modeling to predict process and material characteristics • Develop in-line measurements and visualization: transfer laboratory analytical methods to in-line

- Developing validated models and fundamental understanding that can accurately predict the performance and reliability of nanomaterials and nanocomposites,
- Advancing methods for in-line multi-dimensional (size, shape, charge, etc.) high throughput nanocharacterization essential for scaleup, uniformity, and consistency (reproducible quality),
- Advancing methods for increasing manufacturing throughput, yield, and quality.

The prioritized results of the workshop are presented in Exhibit E-2, including major barriers and solutions in each of the topic areas.

A concerted effort by industry, government and others to address common issues would expedite the resolution of the challenges listed in Table E.2. One important idea that emerged at the workshop was the development of an umbrella alliance which is a multi-industry Nanomanufacturing Consortium, one that would be industry driven (industrial decision makers with long-term commitment, trade associations, and individual companies) and government facilitated and supported. The purpose of this Consortium would be to help consolidate multi-industry research needs, map existing research and capabilities, consortia, and initiatives, and build a nanomanufacturing knowledge base in targeted areas. Moreover, the Consortium would accelerate progress in nanomanufacturing science, engineering, and technology in areas of common need through identification and communication of common needs, and coordinate interactions with funding agencies, Federal laboratories, and academic researchers.

While there is both considerable Federal and private investment in nanotechnology research, the challenges remain in prioritizing what needs to be done and leveraging outcomes (gaining the most benefit for dollars spent). Many government agencies are making a push toward cross-industry programs to expand the reach of research and development. For the U.S. this coordination is critical to attaining and maintaining global leadership in a competitive environment and to achieving the benefits of nanomaterials and nanotechnologies.

1. INTRODUCTION

Nanotechnology holds the potential to remove long-standing limitations, solve technical problems, and even revolutionize manufacturing processes and products across many industry sectors, such as medical technology, electronics, aerospace, automotive, chemical, food and dietary supplements, forest products, and pharmaceuticals. Global investment in nanotechnology rose to nearly $13.5 billion in 2007, with a global market predicted by Lux Research to grow from $147 billion in 2007 (see Exhibit 1-a) to several trillion dollars in the next decade. But significant technical barriers inhibit the discovery, high-volume manufacture and widespread commercialization of nanotechnology-enabled products needed to fuel this growth. The United States Government has invested over $10 billion over the last ten years and is a long-standing leader in nanotechnology funding. Other nations are now rapidly increasing their investments and eroding this lead.

Other parts of the world are emphasizing applied research and development (R&D) to accelerate commercial development of nanotechnologies. The U.S. must do the same to capitalize on its vast investment, ensure cost-effective manufacturing of new nanotechnology-based products, and maintain global leadership in this emerging field.

To achieve the potential of nanotechnology, viable methods are needed to scale up the production of promising nanostructures to a commercial level without losing their unique and valuable properties. Integrating nanotechnology at production scales often entails resolving fundamental issues of physics, chemistry, biology and other core sciences. A cross-industry approach may be the best way to solve these fundamental challenges. While each industry faces its own set of challenges in harnessing the potential benefits of nanotechnology, many issues currently hindering widespread industrial applications are shared by all. A concerted effort by industry, government and others to address common issues could expedite the pace of nanotechnology advancement and benefit all stakeholders.

In the history of industrial innovation, advances that are sufficiently new and disruptive, i.e., that have the potential to change the basis of competition, present challenges to the inventors and early-adopters. Among these challenges is the high up-front investment in R&D before the added-value can be demonstrated, high capital investment for manufacturing, uncertain or non-existent supply chains, and insufficient scientific foundation and measurement capabilities to understand and measure the factors that need to be controlled to ensure quality and performance of the final product. These challenges were seen in the early days of the transistor and integrated circuits, for example, and are now being faced by a wide range of industries considering nanotechnology-enabled technology. In the current global market, companies must find the means to address these barriers to determine if nanotechnology-enabled processes and products can lower their costs, add value, or improve performance to provide competitive advantage over current technology.

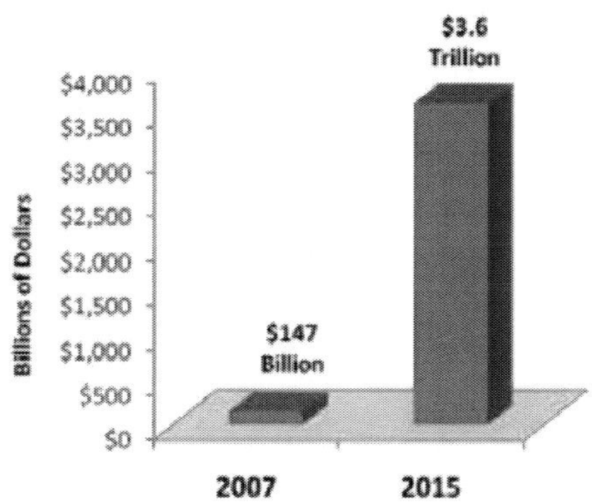

Source: Lux Research

Exhibit 1-a. Projected Economic Value for Nanotechnology Products

1.1. Workshop Overview

The *Cross-Industry Issues in Nanomanufacturing Workshop* was held on May 20-22, 2008 in Gaithersburg, Maryland to explore critical cross-industry issues in nanomanufacturing. The workshop was cosponsoredby NIST, the Agenda 2020 Technology Alliance of AF&PA, the University of Maryland Department of Materials Science and Engineering, Imerys, SAPPI Fine Paper, and USFS. The workshop was coordinated with ongoing activities in "Predictive Modeling of Nanomaterial Properties" addressed in a recent workshop sponsored by the National Science Foundation and the Network for Computational Nanotechnology where it was noted that experimental methods are not always able to probe the length and time scales accessible to electronic structure or simulations at the atomic level.

The objective of the Cross-Industry Issues in Nanomanufacturing Workshop was to identify common problems and solutions specific to nanotechnology, manufacturing processes, and performance of nanomaterials in commercial products. Workshop attendees included experts in the field of nanotechnology manufacturing, as well as generalists in nanotechnology and related fields. A complete list of workshop participants and contributors to this chapter is provided in Appendix A. The workshop included plenary presentations and focused topical presentations throughout the first two days, interspersed with concurrent breakout sessions. The agenda for the workshop is provided in Appendix B; a listof presentations (many of which are available at the workshopwebsite: (http://www ncn. purdue.edu/Content/Workshops/Predictive_Modeling_of_Nanomaterial_Properties_-_Modeling_Needs) is included in Appendix C.

Exhibit 1-b. Description of Cross-Industry Nanomanufacturing Workshop Breakout Topics

Surface/Interfaces and Non-Bonded Interactions of Nanomaterials	Nanotechnology-enabled Composites and Matrices	Separations and Fractionation
Identify fundamental science needs and measurement capabilities required to understand and control the surface-dependent properties of nanomaterials such as dispersion, aggregation, and adhesion at their interface with a matrix, with an emphasis on non-covalent bonding interactions, to enable materials by design and overcome challenges in commercial production.	Identify fundamental science needs and measurement capabilities required to understand and control multiple properties of nanocomposites by design to enable superior performance and reduce cost.	Identify fundamental science needs and measurement capabilities required to characterize nanomaterials and enhance their separation and fractionation to address challenges in commercial production of uniform, high quality, stable, and consistent (reproducible) nanomaterials in high volume and at high throughput.

Breakout Session Topics. Theorganizing committee had worked for over a year, holding small workshops and discussions with stakeholders across industrial sectors, to identify the priorities for crosscutting issues in nanomanufacturing. It's been generally understood in the nanomanufacturing communitythat companies must find the ways and means to determine if nanotechnology-enabled processes and products can lower their costs, add value, or improve performance to provide competitive advantage over current technology. This workshop provided a venue for such discussion. The approach was to find similar problems and potential solutions within widely different industries.Additional input was taken from the

Joint Chemical & Semiconductor Industry Research Needs for Modeling of Nanomaterials, NIST, June 2006; Interagency Working Group Workshop on Nanomanufacturing, NIST, October 2006-http://sites.energetics.com/nanocrosscutmay08/. These priority topics were selected and adopted as the main focus areas for this workshop. Exhibit 1-b summarizes the scope and context of each of the topics discussed. Within each of these topics, common issues included—although were not limited to—measurement, characterization, modeling, performance properties, and environment health and safety concerns. Throughout the workshop, emphasis was placed on identifying the needs that are cross-cutting in nature and could impact multiple industries and products.

Although not all-inclusive, the common issues shown in Exhibit 1-c served as central themes in all groups. These themes represent some of the known major challenges that must be addressed to ensure that the full potential of nanotechnology can beachieved.

EXHIBIT 1-C. COMMON ISSUES FOR NANOMANUFACTURING

Measurementand Characterization: determine the physical, chemical, biological, and technological properties and distribution of nanomaterials in isolation and in matrices with simultaneous high resolution, statistical significance, and high-throughput capacity

Process Control: control the manipulation and interactions of matter at the nanoscale to create nanostructures with unique properties; control their distribution and properties in the final product

Modeling: develop the science to predict the multiple properties of nanomaterials in complex matricesand establish nanomaterials property data infrastructure and an informatics/knowledge base

Performance Properties**:** develop nanomaterials with high value properties that can't be achieved without the novel properties of the nanotechnologycomponents, or lower costs

Environment, Health and Safety**:** Gain knowledge of factors controlling biological interactions of nanoparticles, free and in matrices. Better enable reliable detection and measurement of the type and amount of engineered and/or preexisting nanoparticles in the workplace and in products across the full product life cycle including end-use environments

1.2. Organization of the Report

This chapter is based on the discussions and recommendations resulting from the Cross-Industry Issues in Nanomanufacturing Workshop held on May 20-22, 2008, and the presentations provided by plenary and other speakers.

The industrial needs are presented in Chapter 2 beginning with a summary section, followed by a section contributed by each industrial sector and national nanomanufacturing center in attendance. This chapter also includes a discussion of life cycle needs.

The remainder of the report is organized around the breakout topics shown in Exhibit 1-b, with a section in Chapter 3 devoted to each. For each breakout topic there is an introduction to the topic, a summary of current scientific and technical advances, discussion of the key barriers to success, and—most important—pathways for priority solutions. Mini-roadmaps of the most important factors for priority solutions (including key R&D needs, timelines, goals, existing solutions, barriers, risks, benefits, and partnership roles) were developed for each. The overall path forward is summarized in the concluding chapter, Chapter 4.

At the end of the report, appendices provide a list of participants and contributors (A), workshop agenda (B), list of presentations (C), and acronyms (D).

2. Industrial Needs in Nanomanufacturing

Generally, nanomanufacturing is defined as manufacturing of products at the nanoscale and targets production of new materials or devices with desired combinations of properties that can't be achieved with conventional methods or materials. The key goals are to use existing manufacturing processing expertise and to combine it with new processes as required. The end product should be new materials that are reliable, safe, and have the desired properties. The figure below illustrates the interdependencies.

There are several benefits and challenges to nanomanufacturing as outlined in this chapter. In this chapter selected industries have identified their individual benefits and challenges. Also provided in this chapter are highlights of supporting efforts within the NSF-funded National Nanomanufacturing Network (NNN). Within this network are connections to various university and national lab centers, projects and experts fromacademic, industrial, and government institutions.

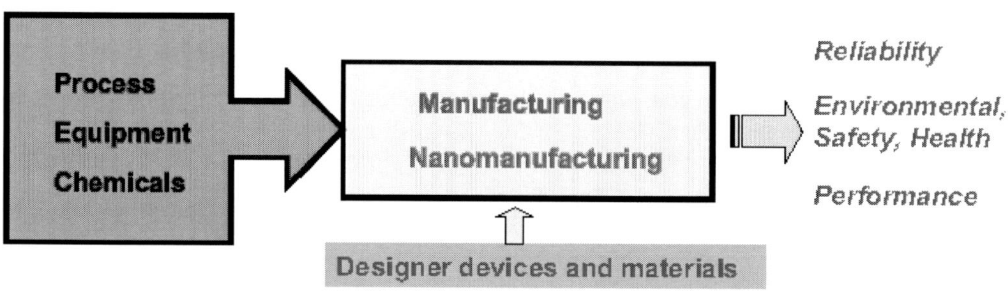

Nanotechnology's impact can be realized only though manufacturing of real products as needed by the different industries. Nanocomposite materials with improved properties can improve the structural integrity of regular materials, expand their operating limits, and provide innovative solutions for many needs throughout practically every industrial segment: electronics, health care, building materials, automotive/transportation, energy technology, and military, to name a few. For example, fundamental needs in the manufacturing and energy sectors for the U.S. include the need to: reduce wear in moving parts; reduce energy consumption; improve the efficiency of alternative energy extraction and utilization from sources such as solar, wind, and bio-fuels; improve catalytic performance of converters or reaction vessels; improve the durability of coatings; and make stronger, lighter composites. These needs can likely only be met through advances in the development of nanocomposite materials. These advances will require the synthesis of new materials and chemicals, the creation of new functional structures, the engineering of new devices, and the application of their use in multiple fields. Moreover, new manufacturing facilities, new technical capabilities, health and environmental studies, regulation for safe use, and adaptation within the educational infrastructure will be required. The synthesis of new materials and chemicals, the creation of new functional structures, the engineering of new devices, and their use in multiple fields will need new manufacturing facilities, new technical capabilities, health and environmental studies, regulation for safe use, and adaptation within the educational infrastructure.

The main challenges that overlap in all the industries represented at the workshop are given in the figure below: bottom-up self-assembly of materials, placement and assembly of materials, predictable and reproducible properties of nanostructured materials. In all the applications, techniques must be developed to (1) integrate nanomaterials into structures where their novel properties are preserved and accessible at the macro-scale, (2) place these structures in predefined locations with sub-nanometer accuracies, and (3) to integrate multiple nanomaterials for achieving desirable system properties. For example, in the semiconductor industry, to be useful in devices, interconnects or package applications, techniques must be developed to integrate these materials into structures where their novel properties are preserved and accessible at the macro-scale, and this is very challenging. For these materials to be useful in devices, the ability to control their nanostructure to enable reproducible semiconductor band-gap must be developed.

For devices and interconnects, the ability to place these structures in predefined locations with sub-nanometer accuracy must be developed. To use these materials in packages in thermal management applications, techniques must be developed to assemble these in high density in predefined directions with a low thermal contact resistance. In the aerospace industry, the additional challenges are two-fold; (1) nanostructured material understanding needs to be communicated and standardized through the entire supply chain, (2) value added compared to current resins needs to be demonstrated.

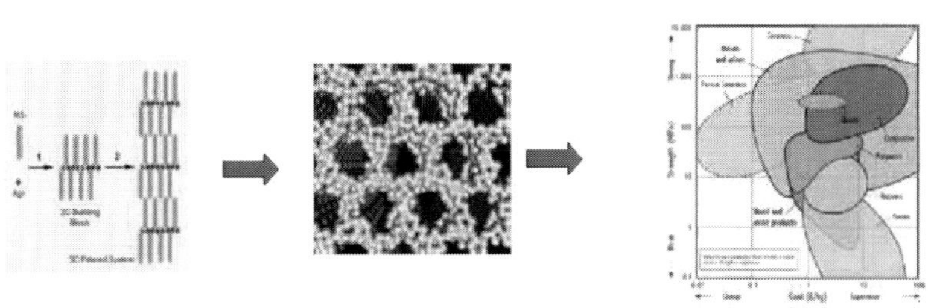

Self-Assembly Placement and Integration Properties

[Images courtesy of S. Shankar, INTEL, Corporation]

2.1. Represented Industries

[Photos courtesy of the USDA]

2.1.1. Food Industry

Food-based nanostructures may be envisioned as "structural composites", where they may provide tailored textural properties and increased shelf stability in foods as well as offering the possibility of food-grade matrices for controlled release applications. Structure formation in all materials, including foods, is a highly complex process. Once molecules self-assemble, they are thermodynamically driven to self-aggregate into larger structures in the ~100 nm range in a process governed by non-covalent molecular interactions (e.g., hydrogen bonding and electrostatic interactions). For proteins, self-assembly drives the formation of secondary, tertiary, and quaternary structures. A similar approach can be envisioned for lipid and carbohydrate self-assembly.

The bioavailability, efficacy, and functionality of food ingredients, additives, and nutraceuticals may be greatly enhanced through nanoscience and nanotechnology. For example, current research involves incorporation of water- and oil-soluble ingredients—all of which are food-grade materials—in microemulsions, solid lipid nanoparticles, and self-assembled dairy peptides. In microemulsion research, the goal is to develop dilutable food-grade systems for controlled delivery of bioactives. One of the biggest challenges is the limited number of food-grade surfactants and co-surfactants available. Most microemulsions are not dilutable, which limits their use in beverages. Solid lipid nanoparticles are oil-soluble compounds in a crystal matrix. They offer increased stability through encapsulation. Potential applications include omega-3 fatty acids and vitamins. In dairy research, the goal is to use nanotechnology to design self-assembled peptide-based nanostructures with dual functions of delivery and bioactivity. Along these lines, emulsion,of plant sterols have been shown to enhance the LDL cholesterol-lowering properties in hamsters, and similar emulsions of gamma tocopherol have been shown to increase bioavailability and anti-inflammatory properties in mice. Self assembly of milk proteins is common in nature, and weak interactions lead to protein folding, micelles, or crystals. Recognizing these phenomena, researchers have been able to create milk protein-based nanotubes through enzyme hydrolysis technology and precise control and manipulation of processing conditions. The presence of calcium may play a role in the initial binding of the proteins. The novel protein structure (i.e., 20 nm diameter nanotube with 12 nm hollow core) provides new functionalities, such as texture, encapsulation, shelf-life stability, viscosity (replaces gelatin), and more.

There is 'natural' nanotechnology at work in the creation of food systems and assemblies, such as casein micelles. This strategy, however, is counter to the food industry approach, where a 'top-down' approach (e.g., emulsification via homogenization) produces thermodynamically unstable structures that must be kinetically-stabilized to achieve a suitable shelf life. With nanostructuring, stable entities are created (a 'bottom-up' approach), shifting the focus from microstructures to nanostructures. Nanoscale structuring of foods offers exciting possibilities for enhancing shelf life, functionality, and controlled release of bioactive compounds. Staling, breakdown of gel structures, and phase separation all contribute to textural changes during the storage of foods. With their inherent thermodynamic stability, self-assembled nanostructures may greatly increase shelf life by slowing down, and ideally preventing, such structural changes. This may be equally applicable to the prevention of off-flavors, which are often due to chemical reactions such as oxidation, or changes in color resulting in browning, which is caused by fat and/or moisture migration. The possibility of re-structuring existing food ingredients into nanoscale matrices that increase the viability of processed foods is one of the greatest strengths of this approach.

A specific example of where nanoscale structuring can play an important role is in preventing syneresis, which is the exudation of an aqueous phase from gels that contract over time (in dairy products, gelled desserts, tofu, etc.), and which reduces shelf life. Industry strategies to counter this problem often rely on the presence of added thickeners (e.g., starches) that bind excess moisture. This common strategy does not fundamentally address this problem (and thus the issue of reduced shelf life). Inspired by research in carbon-based nanostructured materials, food-based nanostructures such as self-assembled proteins may be used to generate stable structures capable of resolving this problem.

Lastly, nanoscale controlled release matrices such as microemulsions and solid lipid nanoparticles offer unique possibilities for increased efficiency in drug/bioactive delivery, the ability to cross biological barriers (e.g., cell walls and the blood-brain barrier), passive targeting and local accumulation in tissues, and increased solubility of drugs or bioactive nutraceutical ingredients. However, few if any food-related commercial applications for such nanoscale matrices exist. This field is still in its infancy, though the technological basis exists (e.g., microemulsions, nanoparticles, etc.). It is likely that the food industry will learn from the massive efforts in nanoscale drug delivery currently taking place in the pharmaceutical field.

These few examples show that there are extensive possibilities for the incorporation of nanoscale and nanotechnology derived matrices in foods, such as improvements in texture, shelf life and the development of food-grade controlled release matrices. However, for this revolution or evolution to occur, there are numerous technological hurdles, issues related to consumer acceptability, and safety concerns that must first be addressed.

2.1.2. Semiconductor Industry

The International Technology Roadmap for Semiconductors (ITRS) identifies the challenges that must be overcome for future technologies to continue increasing density and performance for the next fifteen years. The semiconductor industry is currently producing integrated circuits with minimum features less than 50 nm through lithographic pattering and etching and deposition of conventional materials. Continuing improvement requires overcoming challenges in the control of nanostructures and the resulting properties. New materials will be needed to enable technologies with sub-10 nm features in high density, devices with high electrical performance, low resistance electrical interconnects, and package and assembly processes that manage stress in the integrated circuit and effectively remove heat.

The self assembly of nanomaterials in periodic patterns at higher density than possible by conventional lithography has potential applications in extending lithography and in fabrication of high power density capacitors. The directed self assembly of block copolymers into lines and round opening structures with sizes <10 nm could be useful if they could be placed in precise locations with ultra low defect densities. Critical research is needed to enable the controlled placement of structures relative to previously generated structures and eliminating defects. For high power density capacitors, electrical conducting elements need to be assembled in high density with dielectrics placed between them. Self assembly has potential to enable this, but serious challenges must be overcome to achieve high capacitance devices.

Nanostructured materials such as carbon nanotubes or graphene have novel properties including high electrical conductivity, high thermal conductivity and high mechanical strength. To be useful in devices, interconnects, or package applications, techniques must be developed to integrate these materials into structures where their novel properties are preserved and accessible at the macro-scale, and this is very challenging. In devices the ability to control nanostructures to enable reproducible semiconductor bandgaps must be developed. Devices and interconnects require the ability to place these structures in predefined locations with sub-nanometer accuracy, which is not currently possible. To use these materials in packages in thermal management applications, techniques must be developed to assemble these in high density in predefined directions with a low thermal contact resistance.

Future integrated circuit technologies need package polymers with multiple properties optimized for application, processing, and use. Nanostructured materials have unique properties, such as low thermal coefficients of expansion, high mechanical strength, high toughness, and unique electrical properties. The ability to integrate these into polymers could enable simultaneous achievement of multiple desirable properties, but there is no "rule of mixtures" for nanocomposites. Similarly, the ability to integrate multiple nanomaterials into a composite, is needed enhance the potential for multiple properties. Therefore all potential interactions of nanomaterials in the composite structure must be fully understood and controlled.

In all applications, the ability to control the nanostructure of materials and their interfaces to other materials is critical to provide solutions to future integrated circuit technology needs.

2.1.3. Forest Products Industry

Nanotechnology is an emerging area that carries enormous promise to revolutionize materials use leading to novel applications and products not feasible with conventional technologies. The forest products industry looks to nanotechnology as a means to tap the enormous undeveloped potential of trees as photochemical "factories" that produce abundant sources of raw materials using sunlight and water. Forest biomass resources provide a key platform for sustainable production of renewable, recyclable, and environmentally-preferable materials to meet the needs of society in the twenty first century. Wood-based lignocellulosic materials (i.e., forest biomass) provide a vast material resource and are geographically dispersed.

The forest products nanotechnology roadmap (www.nanotechforest.org) identifies the industry vision as "*sustainably meeting the needs of present and future generations for wood-based materials and products by applying nanotechnology science and engineering to efficiently and effectively capture the entire range of values that wood-based lignocellulosic materials are capable of providing*". In addition, the forest products industry sees its inherent strengths to include stewardship of an abundant, renewable, and sustainable raw material base; a manufacturing infrastructure that can process wood resources into a wide variety of consumer products; and being uniquely positioned to move into new, growth markets centered on bio-based environmentally-preferable products. Nanotechnology will enhance industry's ability to produce new high performance consumer products from lignocellulosic-based materials. The industry vision is well aligned with society's need for establishing a source of sustainable materials and products. The following are priority areas for nanotechnology in the forest product industry.

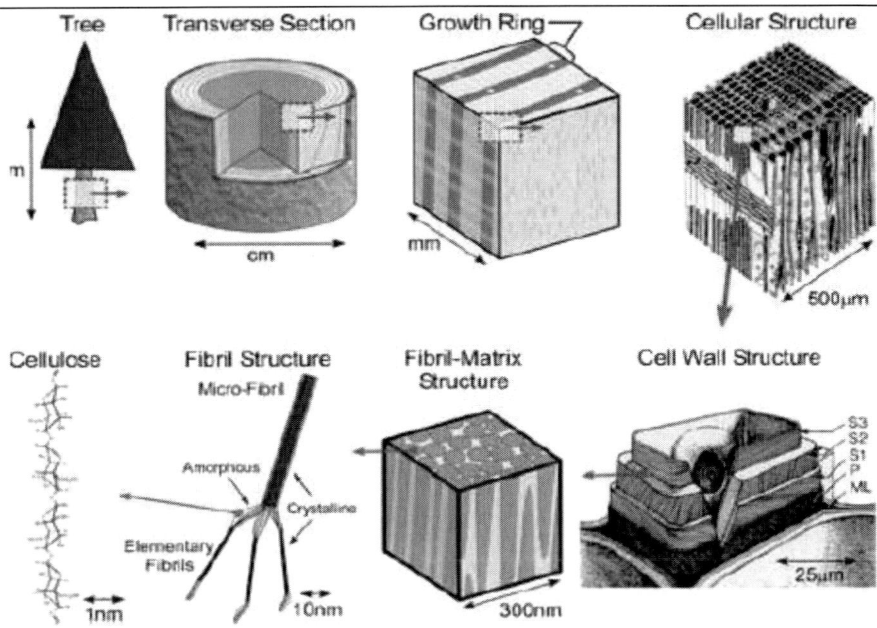

Wood Nanodimensional Structure and Composition

This figure depicts how wood is a cellular hierarchical biocomposite made up of cellulose, hemicellulose, lignin, extractives, and trace elements. At the nanoscale level, wood is a cellulosic fibrillar composite. Wood is approximately 30 – 40 percent cellulose by weight with about half of the cellulose in nanocrystalline form and half in amorphous form. Nanocrystalline cellulose is relatively uniform in diameter and length and these dimensions vary with plant species. Cellulose is the most common organic polymer in the world representing about 1.5 x 10^{12} tons of the total annual biomass production. Cellulose is expressed from enzyme rosettes as 3 - 5 nm diameterfibrils that aggregate into larger microfibrils up to 20 nm in diameter. These fibrils self assemble in a manner similar to liquid crystals leading to nanodimensional and larger structures seen in typical plant cell walls. The theoretical modulus of a cellulose molecule is around 250 GPa, but measurements for the stiffness of cellulose in the cell wall are around 130 GPa. This means that cellulose is a high-performance material comparable with the best fibers technology can produce.

[Sidebar and graphics courtesy of T. Wegner, U.S. Forest Service]

Improving the Strength to Weight Performance. The strength of lignocellulose-based fiber networks in paper and board are controlled by matrix components, bondingstrength, fiber strength, fiber size and shape, effects of any additives or fillers, and uniformity of material distribution. In addition, there are end use product requirements for other key properties such as optical performance, surface smoothness, and stiffness. Large increases in strength-to-weight performance are not attainable with current technology.

Liberation and use of nanocellulose and nanofibrils. At the nanoscale, wood is composed of elementary nanofibrils (whiskers) which have cross-section dimensions of about 3 – 5 nm and are composed of cellulose polymer chains arranged in ordered (crystalline) and less ordered (amorphous) regions. Nanocrystalline cellulose has strength properties approximately equivalent to Kevlar®, and the forest products industry focus is on liberation and use of both nanofibrils (amorphous and crystalline cellulose) and nanocrystalline cellulose. Success will allow the forest products industry to be a major supplier of nanoparticles for a wide range of other industries. Because of the hundreds of millions of tons of wood available for processing, commercial production would be both sustainable and renewable as well as create an industrially significant supply. High value, renewable nanotechnology-enabled composites can be produced by identifying commercially attractive methods to liberate both nanofibrils and nanocrystalline cellulose and by establishing methods for characterization, stabilization, and blending of these wood-based nanomaterials with a variety of other nanomaterials.

Understanding water-lignocellulosic interactions. The response of wood and wood-based materials to moisture and their durability under high moisture conditions is due almost entirely to the super molecular structure of constituent biopolymers (i.e., cellulose, hemicelluloses and lignin) and nanoscale structures. An understanding of the interactions among moisture and woody material components at the nanoscale will enable development of new, innovative technologies that will improve both the durability and dimensional stability of wood and wood-based materials. In addition, understanding and manipulating the interactions between water and wood/paper should permit huge reductions in energy and water usage in converting wood to consumer products and allow more effective and efficient use of wood raw materials in a broad base of new and existing products. Lastly, development of a substantial knowledge base of lignocellulosic/water interactions is expected to lead to the development of new and improved products through control/modification of surfaces to provide barriers to prevent or control the transfer of moisture. This will enable the substitution of sustainable wood-based products for some products derived from less sustainable materials.

High-performance nanotechnology-cellulosic enabled composites. Lignocellulose in woody plants is one of nature's most abundant materials, and wood-based lignocellulose at the macroscale level is one of our most used and ubiquitous materials. To date, the intrinsic self-assembling nanoscale structure of lignocellulose, as well as the versatility of its constitutive biopolymers, has not been fully exploited. Cellulose is a material that has unique tensile properties. However, during the processes to extract cellulose from trees or plants while removing lignin, the length and strength of the cellulose fibers are significantly decreased. It is desirable to achieve the next generation of high performance composites in which cellulosic nanomaterials provide maximum tensile strength and to use cellulose and cellulose-derived nanomaterials with varying combinations of other organic and inorganic materials to produce high-performance, multifunctional composites tailored to specific end-use performance needs. Potential composite properties besides strength include formability and geometrical complexity at the nanoscale, resulting in other unique physical properties, surface smoothness, biomedical compatibility, and ability to reinforce polymer foams.

Papermaking: Self-assembly at 60 mph

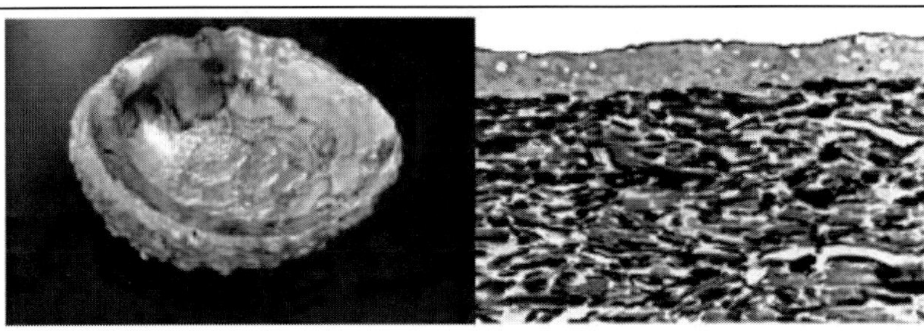

Coating Structure

The structure of clay coating on paper is similar to the structure of an abalone shell (upper left), which achieves its remarkable strength and optical properties from a matrix consisting of 98% $CaCO_3$ nanoplatelets and 2% protein. The top image at right depicts a printed, coated paper. The thin top layer is ink, while the layer under the ink is the clay coating, designed to hold out the ink in a manner that maximizes print quality and performance. The bottom layer is the paper base, consisting largely of treated wood fibers and some pigments.

Nanotechnology-photonics. Many grades of paper require using higher grammage (basis weights) than needed, not because of strength property end-use requirements but because of the need to achieve sufficient opacity. While it is desirable toachieve costs and materials savings by reducing the amount of raw material for a given unit of functionality in both fiber and coating, optical performance of the paper cannot be compromised. More efficient optical performance with minimal weight will benefit all grade levels but especially the ultra-lightweight grades where opacity decreases rapidly with weight. Benefits include significant reduction in materials, processing and distribution costs. In order to be able to make high opacity coated paper in the same weight range of tissue paper, combinations with nanomaterials such as graphene, carbon nanotubes for electronics, optic manipulating nanomaterials, nanomaterials for high strength applications, and nanomaterials to control ink interaction must be explored.

2.1.4. Chemical Industry

The chemical and materials science industries are ideally suited to exploit the benefits of nanotechnology. The synthesis of nanoparticles, and their incorporation into products, is squarely within the scope of the activities of this industrial segment. The chemical engineering know-how in this sector puts the sector in a leading position to bring the potential of these new materials to fruition.

The potential benefits of nanotechnology, to the business enterprise and to society in general, are colossal. The creation of nanocomposite materials with improved properties can improve the structural integrity of materials, enhance their environmental operating limits, and provide innovative solutions for many needs throughout practically every industrial segment: electronics, health care, building materials, automotive/transportation, energy technology, and military, to name a few. Fundamental aspects of the operation of our society, such as: the need to reduce wear in moving parts; improving catalytic performance of converters or reaction vessels; improving the durability of coatings; extending control over drug delivery; making stronger, lighter composites; improving the efficiency of alternative energy sources such as solar, wind, and biofuels; all can be addressed with nanotechnology.

The integrated impact of nanotechnology will fuel economic growth for the next few generations. Nanotechnology may account for millions of jobs in the near future. The synthesis of new materials, the creation of new functional structures, the engineering of new devices, their implementation throughout society—all these activities will necessitate new manufacturing facilities, new technical capabilities, health and environmental studies, regulation for safe use, and adaptation within the educational infrastructure.

Realizing this potential means overcoming the challenges that we presently face. These challenges are as diverse as the resulting benefits. On a basic level, we do not yet have sufficient knowledge about the nanomaterials to be able to predict the wealth of improvement that can be made. Quantifying the magnitude of potential improvements is an ongoing effort in industry and academia at this moment. Nanomaterials push against our metrology limits. Characterizing their physical properties has already created a suite of new experimental approaches, and these have, in turn, begun to spawn new technologies. Manufacturing technology needs to be developed to generate the tons of material that will be necessary. We need to understand how these materials will co-exist with the ecological web of life on the planet. Health risks must be mitigated to be commensurate with potential rewards. There is also a challenge in education—preparing students to participate as workers in this field.

Research needs to be incorporated into the current paradigm, and then disseminated to new generations of students, so that nanotechnology will be established at universities ten years from now in the same way molecular biology was established twenty years ago.

Finally, a clear national strategy needs to be enunciated. Currently, nanotechnology is in an embryonic stage. It is important to learn from the lessons of the past—chemical manufacturing, molecular biology, nuclear energy—to create a strategy that will bring lasting benefits to the world. The chemical industry has proposed a plan for a roadmap through 2020 in which the following needs were identified; specifically on manufacturing and processing. Several unit operations have been identified by the chemical industry as critical areas of focus; synthesis, separation, purification, stabilization, and assembly. In addition, surface engineering needs to address catalysts, ceramics, coatings, sorbents and membranes. Few programs are dedicated to issues of nanomanufacturing, and in the existing programs scale-up issues are not fully addressed. A research center or virtual center dedicated to process development is needed to provide sufficient impetus for progress in this area.

2.1.5. Aerospace Industry

In the aerospace industry, nanotechnology is seen as one possible route to the material systems and components of tomorrow's aircraft. The tremendous promise of nanotechnology to provide breakthrough material solutions has generated wide interest in nanotechnology-enabled materials among aerospace industry researchers. Material needs are as varied as their applications, and nanotechnology-enhanced materials provide opportunities to fill many of these needs. Aerospace structural material applications require materials that are large in scale, often meters in one or more dimension, such that the nanotechnology-constituents provide improvements to macroscopic materials properties. Aircraft electrical components, on the other hand, benefit more directly from nanotechnology for miniaturization, resulting in decreased power demand, lighter weight, and in some cases increased sensitivity. Such components are developed by the semiconductor industry; section 2.1.2 contains a detailed discussion of the electronics industry's needs in nanomanufacturing. The remainder of this assessment focuses on structural material systems.

The impacts of nanotechnology on the aerospace industry are broken into two categories: those technologies that allow the aerospace industry to improve its current designs, and those technologies that open new regions of design space. A nanomaterial that offers an incremental cost or weight savings over a current material, a nanotechnology-enabled electronics box that weighs less and uses less power, or a nanoscale surface modification to improve environmental protection, allow improvements in future designs. Weight savings are particularly precious to aircraft design; any material weight savings is realized several times over once the reduced propulsion, lift, and fuel needs are considered.

Other technologies may offer disruptive changes to the industry. Nanotechnology-enabled distributed sensor systems are being researched for structural health management, which may extend the usable lifetime of a vehicle. Lighter and stronger structural materials, especially with improved ablative properties, could extend the range, speed, and maneuverability limits of fighter aircraft. Structural materials that incorporate microwave absorption would provide increased defense against electronic attacks. Nanotechnology has the potential to enable multiple functions in a single material, which will be a great boon to aerospace design.

The aerospace industry has stringent certification requirements for materials to be used in planes and spacecraft. The high level of quality and process control applied to materials for the aerospace industry has caused certain materials to be classified as "aerospace grade," where materials without that qualification are considered "commercial grade." The methods used for certification drive an intense need for repeatability in material performance, which in turn drives a need for repeatability in the synthesis of materials. Manufacturing and processing technologies have been developed to produce consistent aerospace grade materials; moving to nanotechnology-enhanced materials will require the development of new manufacturing technologies and new processing technologies. In-process monitoring and control of dispersions and non-destructive proof of proper dispersion are nanotechnology needs that will enable the certification and use of nanomaterials. Other technologies are required to produce certifiable nanomaterials besides dispersions, such as surface roughness measurements over square meters in ambient conditions to certify surface treatments.

The ability to determine bulk properties of nanomaterials based on their constituents, or even better the ability to design a nanomaterial based on the required bulk properties, lies at the end of a very long road of future research. The understanding required to predict bulk properties and produce consistent nanomaterials is being gained through research via experiments and computational modeling. Establishing research and reporting standards and applying informatics will hasten the realization of these goals.

The aerospace industry has a wide variety of needs to which nanotechnology is beginning to be applied. Discovery of new solutions can be accelerated by understanding how to surpass or enhance existing materials in highly repeatable nanotechnology-enhanced material syntheses.

2.1.6. Filtration and Separation Industry

Nanoscale needs manifest themselves in multiple areas for the filtration and separation industry. There are specific products that need to be either purified or separated for the end user in both the industrial and life sciences market areas, and which can only be processed using filtration. The materials that are used to enable the filtration and separation require finer and finer elements down to the nanoscale in order to meet these stringent demands.

The life sciences market encompasses biopharmaceutical, biotechnology, and medical requirements. Manufacturers in the pharmaceutical and biotechnology areas require certain products to have viruses removed, contaminant particulate removed, and final product selected down to a set level in the nanotechnology region. Filtration and separation materials for such applications must have the correct selection capabilities and low extractables, meet government requirements in regards to use, and have high throughput and flow/flux capabilities. The challenge is being able to do this at an affordable cost. To accomplish this, quite often a series of separation steps are used in conjunction, for example, pre-filtration, chromatography and final filtration.

The life sciences market also covers diagnostics and detection where concentration, selection, and detection of molecules in the nanogram, or below, concentration range are to be found for various disease states and wellness programs. Obviously as the amount of material is low to start with, the amount of sample lost on route to the final detection point should be minimized throughout the process.

Medical requirements can be fairly stringent as nanoparticulates can be used for targeting tumors and drug delivery applications. All materials must meet regulatory as well as the application's needs.

The filtration and separation industry markets encompass food and beverages, semiconductors, power, water, chemical, transportation, aerospace, and military. Quite often the filtration and separation employed here is required to remove unwanted side products or contaminants from a production process making materials that will be used in a consumer product. Many of the filtration products deployed in these markets, combine polymeric and inorganic porous materials that have structures on the nanometer scale.

Nanoscale organic polymers and inorganic materials need to be essentially pure, because unwanted larger particles within a nanotechnology product can have undesired effects in any composite material made from them—such as mechanical failure, non-uniformity defects, or adverse pharmacological effects.

Filters are used to remove a wide range of submicron, and nanoscale particulate matter and microorganisms, which would otherwise have an adverse effect on the product quality. In some cases, the end product is entirely enabled by the filtration process. Further still, filtration technologies can be deployed to remove soluble species to purify a fluid, for example, contaminants such as metals, or salts. This application is truly "sub-nano" in scale and requires filtration and separation structures with very fine pores.

With the drive in nanoscale processes for faster, more efficient ways to achieve the required end product purity level whilst maintaining affordability, the combination of multiple filtration and separation steps with on line nanoscale detection could help to achieve these lofty goals.

2.1.7. Industrial Minerals Industry

Industrial minerals are widely used in diverse applications and are the fabric of society. Many applications, such as cement/concrete, ceramics, filtration materials, bricks, plastics, paint/coatings, and paper coatings and fillers, have been in use for centuries where mankind has empirically developed formulations and methods of manufacture. Largely, these have been seen as traditional industries with no new scientific challenges. In fact these materials and applications are so complex that until recently people have largely stayed away from studying them at a fundamental level. Developments in analytical techniques are now allowing us to begin to understand the nature of these materials and realize that many of the key performance features are the result of interactions taking place at the nanoscale. In particular, studies of materials such as abalone shells, bone, dental enamel, and sea sponges are pointing the way to manage the interfaces and develop high strength materials through these biomimetic pathways.

For many years bentonites or montmorillonites have been intercalated and surface modified to provide nanocomposites of remarkable mechanical performance. These represent "nano-minerals" that exist relatively abundantly in nature and require modest processing to refine them to usable performance specifications. Other minerals such as calcium carbonate, alumina, and diatomaceous earth can be ground to nano-dimensions that expand their applicability. Many of these natural mineral forms can also be synthesized at large scale. Precipitated calcium carbonate (PCC) is just one example of this 'bottom up' approach to nano-minerals and is currently manufactured with primary particle size in the 20-70 nm range.

Mining abundant nanoparticles(Image courtesy of P. Jones, Imerys)

70 nm PCC for high impact, low temperature ductility PVC

Properties by Design: Synthetic mineral nanoparticles tailored to improve composite properties

"Synthetic minerals" refers to synthetic approaches to inorganic materials that allow control of product morphology while retaining the relatively low cost basis of common mineral compositions. These synthetic approaches are 'bottom-up' (growth from molecular scale) rather than 'top-down' (size reduction). Precipitated calcium carbonate (PCC) is made this way from lime and carbon dioxide. In nanoscopic form, it is used in plastic reinforcement, in rheological improvement of inks and sealants, and in high quality coated ink jet papers. The advantage of synthetic minerals often lies in their characteristic narrowness of distribution in dimension and shape. This allows for tuning of their morphology to the needs of a specific application. A typical application for many years is in highly impact-resistant PVC fittings such as the pipes and joints shown in the figure above.

The challenge for synthesis is control of primary size, shape, surface chemistry, and dispersion. It is also possible to find minerals existing in the ground with at least one dimension less than 100 nm and showing performance properties different from the same minerals that have all dimensions greater than 100 nm. These "nano-minerals" such as thin crystal kaolin and halloysite have, for the most part, been overlooked because they have shown such radically different properties from commonly used clays and often have been difficult to disperse and stabilize. It is expected that there are many more nanodimensional minerals available for mining. We need the necessary protocols to be established to allow their discovery.

The opportunity exists to increase exploration work to identify other nano-dimensional minerals that will be cheaper to recover and beneficiate so that we can expand their use into commodity applications. The existing minerals industryprocess in the United States has substantial capacity in place not only to mine but also refine minerals by "cracking" (grinding) and "distilling" (by a variety of centrifuge types) nano-dimensioned minerals. Similarly it is possible to capitalize on the installed capacities to synthesize minerals. Often these nano-dimensioned minerals have unusual properties compared with conventional mineral forms and have been passed over because of this. The opportunity exists to focus efforts and tools of nanotechnology to enable the manufacture of large volumes of well characterized nano-minerals at price points of less than $1 per pound.

As with other materials discussed in this chapter, the use of industrial minerals is challenged by the same lack of understanding of means to separate and stabilize nanotechnology mineral components, the ability to disperse and handle these building blocks, and to assemble them into useful composite structures. Findings emerging from biomimetic studies are beginning to point the way towards ways of assembling minerals into tougher pragmatic composite materials[1]. Combining natural and synthetic approaches to develop hierarchical composite nanostructures may be the ultimate way to bring the advantages of the nanoscale into practical, large-volume applications.

A number of formidable scientific and technical challenges remain with respect to advancing nanotechnology involving the production and use of nano-dimensional minerals. These challenges include characterization, separation and refining, dispersion, surface chemistry and interface interactions, and self-assembly. Nanoscale minerals need to be able to be quickly and accurately characterized with respect to their particle shape, particle size distribution, and bulk and surface chemistry. Commercially viable processes are need for separation and refining based upon mineral type, mineral particle size, and mineral three-dimensional shape. New and improved dispersant technologies are needed to effectively and efficiently disperse nano-dimensional minerals in order to keep the particles separate. In addition, it is also desirable for dispersants to be functional so that they can participate in subsequent reactions. Additionally it is critically important to be able to understand the surface chemistries of nano-dimensional minerals and to be able to quickly characterize them with respect to bonding site types, location, and bonding site density. It is also important to be able to identify the range of chemistries useful for a particular mineral species. Lastly, it is important to identify and develop self-assembly strategies that will allow production of high-strength and high-toughness composites made up of varying combinations of minerals and other organic and inorganic constitutive materials and that enable development of smart surfaces and structures.

"Synthetic minerals" refers to synthetic approaches to production of inorganic materials that allow control of product morphology while retaining the relatively low cost basis of common mineral compositions. These synthetic approaches are "bottom-up" (growth from molecular scale) rather than "top-down" (size reduction). PCC is made this way from lime and carbon dioxide. In nanoscopic form, it is used in plastic reinforcement, in rheological improvement of inks and sealants, and in high quality coated ink jet papers. The advantage of synthetic minerals often lies in their characteristic narrowness of distribution in dimension and shape. This allows for tuning of their morphology to the needs of a specific application. A typical application for many years has been in highly impact-resistant PVC fittings such as the pipes and joints shown in the sidebar above.

2.1.8. Pharmaceutical and Medical Device Industries

The pharmaceutical and medical device industries are aggressively pursuing "nanomedicine" applications of nanotechnology and nanoparticles.[2,3] Pharmaceutical applications of nanotechnology include drug discovery & development, drug and gene delivery, nanoparticle drugs, and *in vivo* imaging and cell labeling. Medical device applications of nanotechnology include *in vitro* diagnostics, biomaterial composites for dental, orthopedic, and tissue engineering scaffolds, and active biomedical implants and artificial organs.

Drug discovery. Nanotechnology is currently being investigated to speed up the drug discovery and development process. Applications include, but are not limited to: (1) nanoparticles for detection, quantification, synthesis and analysis of drug targets; (2) rapid target screening; (3) rapid target validation and assay development with reduced sample and reagent volumes and cost; (4) nanofluidics; and (5) personalized medicine approaches that encompass imaging studies, rapid screening, rapid optimal formulation development, toxicity studies using remote sensing, and the identification and measurement of compounds that are indicative of disease.

Drug delivery. Nanoparticle drug delivery mechanisms and formulations can improve therapeutic efficacy and bioavailability. Issues such as poor solubility, tissue damage, rapid breakdown of drugs *in vivo*, unfavorable kinetics, poor biodistribution, and lack of selectivity are being solved with applications of complex nanoparticle surface and layered structures.[4,5] The development of controlled release nanoparticles appears to be a rich field for drug delivery improvements. Nanoparticle properties that may impact drug delivery include size, shape, charge, hydrophobicity, and surface targeting (labeling).

Drugs. Nanoparticles themselves may be used as therapy in the treatment of disease. Applications include dendrimers and fullerenes.

In vivo imaging. Nanoparticle MRI and ultrasound contrast agents have been developed, and much current work is being done to develop extra- and intra-cellular optical and infrared reporters taken up by specific cell types and sensitive to individual molecules within the cells.

Diagnostics. Nanoparticle-based diagnostics promise to dramatically improve the sensitivity and specificity of detection of biological molecules, enabling measurement of compounds that are rare (i.e., outside the range of compounds measured in currently available medical diagnostic tests) or that are present at only minute levels in the body. Current approaches make use of *in vitro* samples for testing, but nanoparticle-based diagnostic approaches are predicted to enable *in vivo* monitoring within the body. In fact, new methods of nanoparticle-based diagnostics are frequently reported in today's literature base. Some examples include carbon nanotube-based sensors,[6] multiplex protein assays using magnetic nanotags with giant magnetoresistance sensing,[7] quantum dots for fluorescent sensing, and gold nanoparticles for SERS (Surface-Enhanced Raman Sensing).

Biomaterials and Implants. Nanotechnology research promises to improve the design and control of material bulk and surface properties. The research is enabling development of better nanocomposite materials for dental and bone replacements, porous materials for tissue and bone regeneration and replacements, surface modification and control, including non-fouling and biofilm-suppressing nanomaterial coatings to improve infection control in medical devices, non-thrombogenic surfaces for cardiovascular implants, nanoporous materials for nanofiltration separations, and nanocoatings for biomolecule adsorption and immobilization, to name but a few examples.

These applications of nanotechnology to the pharmaceutical and medical device space will require robust manufacturing processes and measurement systems to meet quality system validation requirements and bring these nanotechnologies to the market.

2.2. Manufacturing and Life Cycle Challenges

The research conducted to date indicates that engineered nanomaterials can undergo rapid alterations in composition, surface property, morphology and physical state. This is especially evident when these compounds interact with other engineered nanomaterials, with other compounds, or as they move through the environment (air, water, soil). Current research also indicates that the toxicity of these materials changes as their properties are changed. For example, a nanomaterialthat has a certain coating that becomes degraded may express toxicity although the original form may not. Consequently, there is aneed to understand where in the life cycle of the product the nanomaterial exposure could occur, what form the nanomaterial is in at that point (detection and characterization), and how stable the resulting nanomaterial is. Research to improve our scientific understanding of how and when these materials enter the environment, how and where they are transported, and what transformations they have undergone, is a critical need for the appropriate and accurate assessment and management of potential risks. An appropriate risk assessment and management strategy is necessary to enable the industry to develop products and materials that will not posemajor risks to human health or the environment.

2.3. Manufacturing Centers

The potential economic and societal benefits of nanotechnology cannot be realized without manufacturing. For the U.S. to excel in manufacturing at the nanoscale, various branches of science, engineering, business, and government must work together.

Nanomanufacturing research and development requires the collaboration of interdisciplinary partners, information exchange, and the integration of diverse manufacturing techniques. The *National Nanomanufacturing Network* (NNN) works to provide connections to nanomanufacturing centers, projects, and experts from academic, industrial, and government institutions through cooperative "real-space" activities and cyberinfrastructure.

2.3.1. National Nanomanufacturing Network and InterNano

In 2007 the National Science Foundation, the NSF Center for Hierarchical Manufacturing (CHM) at the University of Massachusetts, and partner centers announced the launch of the ***National Nanomanufacturing Network***—a community-driven open-access network that facilitates cooperative activities and disseminates information among the nanomanufacturing research, education, and development community. This network serves as an important catalyst for the advancement of new approaches in nanomanufacturing in the U.S. The NNN is funded by the National Science Foundation, as part of the National Nanotechnology Initiative. Further information regarding objectives, partners, and working with the NNN can be found at www.nanomanufacturing.org.

The NNN offers a network of expertise and technologies, thematic workshops on emergent nanomanufacturing methods, educational opportunities in nanomanufacturing and a web-based nanomanufacturing information clearinghouse. The NNN nanomanufacturing information clearinghouse, entitled InterNano (www.internano.org), is a community knowledgebase that is designed to provide information on nanomanufacturing centers, experts and resources, nanomanufacturing processes, nanostructured materials, best practices, events, and a database of nanomanufacturing research information. The NNN links the four NSF nanomanufacturing NSECs: The *Center for Hierarchical Manufacturing at the University* of Massachusetts Amherst, the *Center for High-Rate Nanomanufacturing* at Northeastern/UMass Lowell/UNH, the *Center for Scalable and Integrated Nanomanufacturing* at Berkeley/UCLA, and the *Center for Nanoscale Chemical-Electrical-Mechanical Manufacturing Systems* at the University of Illinois. The NNN also highlights the activities of other affiliated research centers with a nanomanufacturing emphasis, including the *Center for Integrated Nanotechnologies* at Sandia National Labs, other Government labs at NIST, DOD, DOE, NIH, NIOSH, and other academic centers. Perhaps most importantly, the NNN engages industry to help identify and communicate manufacturing needs and challenges to academic and government research centers to develop effective strategies to build a robust platform of U.S. nanomanufacturing. The NNN collaborates with industrial consortia, professional societies, and individual companies on focused issues and roadmapping to aid the advance of nanomanufacturing.

InterNano (www.internano.org) is designed to serve as the premier information service for the nanomanufacturing community and the informatics portal of the National Nanomanufacturing Network. InterNano provides its users with searchable collections on the state of practice in nanomanufacturing that are both current and comprehensive, as well as industry news, networking opportunities, best practices, and research support. InterNano has a long-term strategy to employ the most appropriate and up-to-date web technologies to enable its users to both extract and contribute information to the InterNano knowledgebase. InterNano incorporates informatics technologies to satisfy the data sharing and analysis needs of NM practitioners. InterNano works cooperatively with complementary informatics initiatives in the U.S. to facilitate data sharing with other groups engaged with aspects of nanomanufacturing and nanotechnology.

Key Needs. Nanomanufacturing process technologies are currently at various levels of maturity, with most at a state of infancy. To effectively evaluate the value proposition for a particular nanomanufacturing process technology a range of issues must be considered. First of all, the process must be built upon robust science and technology so that the impact on product performance is significant. This involves an accurate determination of mean value and statistical distribution of each physical property associated with the target nanomaterials, as well as the reproducibility and reliability to produce those values. Without this information it is difficult to design products and a manufacturing process to produce them. Huge uncertainty in design margins presents too great of a risk to a manufacturer. Guaranteed reproducible physical properties requires reliable process metrology, calibration standards, materials or component certification, documentary standards, and ideally, a deep knowledge of manufacturing process-property relationships. For the last, the more that can be understood and modeled in terms of fundamental science, the better, although in manufacturing practice this is not always achieved.

Currently nanomanufacturing is involved in perhaps only one or two steps in the value chain of a product's production, perhaps as a nanomaterial feedstock or as a value-added nanostructuring process. In integrated process situations, a nanomanufacturing process needs to be compatible with the surrounding process technologies, used prior to or subsequent to the nanomanufacturing process step. This means that the physical or chemical conditions of one process step should not deteriorate or hinder the properties achieved by another—or to detrimentally contaminate the process equipment. In the future, it can be envisioned that many more nanomanufacturing processes may involved, either as sequential processing steps, or in simultaneous guided self-assembly schemes, resulting in a heterogeneously integrated nanosystem.

Availability of the requisite nanomanufacturing process tools is another important consideration. Although some companies are vigorously developing process tools for nanomanufacturing, there are plenty of opportunities to develop commercial-scale equipment for processes that have demonstrated proof of principle, but currently only exist at the laboratory level. The decision to purchase nanomanufacturing equipment, at some cost, is weighted by its extensibility to future manufacturing opportunities. That is, the equipment is of the greatest value if it can be effectively deployed in one mode of manufacturing or another for an extended foreseeable future. If it is to become obsolete quickly, it has little competitive advantage. The decision between performing a manufacturing process step in-house versus outsourcing is, as always, another factor.

New process technologies require skilled operators and engineers. Workforce training is an important issue. As new nanomanufacturing processes develop and mature, education and training is needed through university, technical community college, and professional certification programs, in which the NNN partner institutions play a critical role. Responsible manufacturing also includes well-planned environmental, health and safety controls throughout the product life cycle. NIOSH, FDA, EPA, and other organizations provide important guidance for worker, consumer, and environmental protection.

The NNN and Nanoinformatics. The nanomanufacturing community requires data, information, and knowledge. Nanoinformatics involves the development of effective mechanisms for collecting, sharing, visualizing, and analyzing information relevant to the nanotechnology. It also involves the utilization of the latest information and communication technologies to support efficient communities of practice. Various organizations in the U.S. are currently working together is to identify and prioritize nanoinformatics needs, discuss ongoing activities, and draw up strategies for the future. InterNano is the informatics arm of the NNN, working to identify, aggregate, analyze, and apply information in the nanomanufacturing domain. In addition to the functions described above, InterNano includes a dynamic taxonomy of nanomanufacturing process terms, to aid the organization of information pertinent to the nanomanufacturing community. The NNN and other organizations focused on complementary nanotechnology subject domains (health, safety, simulation, and others) have identified the need to create a system of federated nanotechnology databases. This is leading to the creation of a pilot project that includes the development of a minimal set of metadata standards that will lead to very effective database federation.

The NNN and International Standards. Nanomanufacturing is a global enterprise. As such there is the strong need for useful international standards. The American National Standards Institute (ANSI) has submitted a proposal to the International Standards Organization (ISO) TC 229 to develop documentary standards on nanomanufacturing process terms and definitions. The standards proposal was modeled after the NNN InterNano taxonomy combined with a British Standards Institute document on nanofabrication terms. This project, slated to begin in 2009 and co-led by the U.S. and the UK, will be an opportunity for a wide range of international companies and organizations to weigh in and reach consensus on an initial list of nanomanufacturing terms and definitions. Other ISO standards with nanomanufacturing relevance are in process or under development. It is vital that U.S. industry have strong representation in these standards development activities, either as direct members of ANSI or indirectly through cooperation with current ANSI members.

2.3.2. National Nanotechnology Manufacturing Center

The National Nanotechnology Manufacturing Center in Georgia (http://nationalnano.info/aboutus.aspx) is a non-profit corporation dedicated to the development of a robust, sustainable, domestic nanomanufacturing infrastructure by leveraging existing manufacturing capabilities and supporting the creation of new capabilities when necessary. To address technological barriers in manufacturing, for example, NNMC would identify small-to-medium sized existing manufacturers as a resource for pilot and initial production runs. Small-to-medium sized manufacturers offer several important advantages over any comparable research facility. These are:

- Knowing how to make small volumes profitably so even the earliest production can generate revenue or at least break-even. This transition from strictly investment to cash generating is a vital milestone in any project.
- Making use of existing manufacturers to employ established process control schemes to the material/device and to provide a robust basis for identifying needed

improvements. As process control improvements are defined, they can be implemented in a system that allows assessment of impact.
- Generating product within an established manufacturing paradigm. This supports validation and accelerated commercialization by prospective customers. There is frequently a reluctance to qualify product for an application if it has been manufactured in a research facility because product from a different facility may have significant variation.

Finally, engaging the innovative spirit as well as the physical capabilities of America's small manufacturers may be the surest way to catalyze an economic revolution along with the technical revolution.

2.3.3. Forest Products Center – University of Tennessee

The University of Tennessee's Forest Products Center conducts multi-disciplinary research and outreach directed at the effective and efficient utilization of renewable resources (http://wood.Tennessee.edu and https://www.utbioenergy.org). The program is structured into two broad science and technology areas. The bio-based materials program, including engineered composites and nanotechnology-reinforced polymer systems, is a longstanding emphasis area for the center. The program encompasses fundamental studies on interfacial structure and properties of composites, as well as development of novel process technologies to yield performance improvements in materials. Recently, the center has expanded to investigate opportunities for biorefining of lignocellulosic biomass. Research in this area focuses on defining fundamental transformation pathways to create value-added chemicals and products from the carbohydrate stream of the biorefinery. Additionally, methods of cell wall deconstruction that enhance carbohydrate release are under development. The Forest Products Center has acquired considerable expertise in rapid assessment methods for the determination of relevant properties, a capability that cuts across the two program areas, and builds on the center's process monitoring strengths. The individual program areas are tightly integrated, leading to extensive collaboration between faculty and a unique capacity to address a range of information needs.

3. Workshop Breakout Session Topics

The workshop was structured to identify common problems and common solutions specific to nanotechnology, manufacturing processes, and performance of nanomaterials in commercial products within widely different industries, including the aerospace, automotive, chemical, food, forest products, medical technology, pharmaceutical, and semiconductor sectors. Breakout sessions provided a venue for participants to focus on the three highest cross-industry priorities identified in previous workshops and meetings as they apply to the design, synthesis, and production of nanotechnology-enabled products. The three breakout session topics were:

- Surface/interfaces and non-bonded interactions of nanomaterials.
- Nanotechnology-enabled composites and matrices.

- Separation and fractionation.

Within each of these topics, common issues included, although were not limited to, measurement, characterization, modeling, performance properties, and environment, health, and safety concerns. Throughout the workshop, emphasis was placed on identifying the needs that were cross-cutting in nature and could impact multiple industries and products.

3.1. Surfaces/Interfaces and Non-Bonded Interactions of Nanomaterials

Material surfaces and interfaces play a crucial role in a variety of chemical, electrical, mechanical, and biological phenomena and applications, cutting across a wide swath of industries, government programs, and university research. Chemical, physical, and biological interactions take place on the surface of a particle. The increased surface-area-to-volume provided by nanoparticles and nanostructured materials leads to a dominant role of surfaces/interfaces in the development and performance of nanomaterials.

Material properties that nanomaterial-enabled surfaces and coatings can impact include: lubrication, adhesion, agglomeration, dispersion, wetting, surface tension, surface chemistry, solubility, permeability, cleanliness, biofilm formation, bioactivity, biocompatibility and toxicity, charge, protein binding, and others. Active research continues to explore the range of surface properties that can be expanded through the use of nanoparticles and nanostructured surfaces.

While the past 25 years have seen tremendous growth in the development of tools and techniques to characterize the composition and structure of bulk surfaces, the corresponding tools and techniques to control and predict the performance at nanoparticle and nanostructured surfaces is just beginning. This performance control and prediction of nanomaterial surface properties is required to enable broad, robust manufacturing of products leveraging nanosurfaces.

A large number of nano-dimensional materials have been developed for diverse applications, demonstrating the effectiveness of understanding materials at this scale. However, substantial difficulties in dispersing and stabilizing nanoparticles have emerged. At this scale—where surface area and number of particles are very large—the building block interactions become significant, causing systems to agglomerate. In addition, when incorporated into systems for applications, interactions are further compounded by this added complexity. The non-covalent forces such as Van der Waals and hydrogen bonding can become very large due to the very high surface area and large number of particles. In many natural materials such as wood and bone,the summation of these forces can result in materials with useful strength properties.

Session Scope

- Identify what surface and interface issues are critical to nanoscale building blocks
- Identify what kinds of measurements, chemistry andmetrology are useful to control the nanotechnology building blocks needed for products
- Provide possible solutions to overcome the challenges and barriers for furthering efforts

Exhibit 3.1.1-a. Examples of Commercial Applications of Nanoparticles*

Nanoparticle component	Application	Indication	Company
Liposomes	Drug delivery	Cancer	Liplasome Pharma (Lyngby, Denmark) Schering-Plough Corp (Kenilworth, NJ)
	Drug delivery	Vaccines: influenza, hepatitis A	Berna Biotech AG (Basel, Switzerland)
	Drug delivery	Fungal infection	Enzon (Bridgewater, NJ), Gilead Science (Foster City, CA)
	Nutrient and bioactive food compounds	Improved bioavailability in the body, improved functionality in food system	AquaNova (Germany)
Dendrimers	Therapeutics	HIV, cancer, ophthalmo-logy, inflammation	Starpharma (Melbourne, Austalia)
Carbon nanotubes	*In vitro* diagnostics	Respiratory function monitoring	Nanomix (Emeryville, CA)
	Imaging	Atomic-force microscopy probe tip	Carbon Nanoprobes Inc (Seattle, WA)
Quantum dots	*In vitro* diagnostics, imaging	Labeling reagents: Western blotting, flow cytometry, biodetection	Evident Technologies (New York, NY), Quantum Dot Corp (Hayward, CA), Nanoco Technologies Ltd (Manchester, UK)
Magnetic nanoparticles	*In vitro* diagnostics	Cancer	Immunicon (Huntingdon Valley, PA)
	Imaging, therapeutics	Liver tumors, cardiovasc-ular disease, anemia	Advanced Magnetics (Cambridge, MA)
	Therapeutics	Cancer	Nanospectra Biosciences Inc (Houston, TX)
Gold nanoparticles	*In vitro* diagnostics	HIV	Amersham/GE (Little Chalfont, UK)
	In vitro diagnostics, imaging	Labeling reagents (PCR, RNA, Western blotting), angiography and kidney function testing	Nanoprobes Inc (Yaphank, NY)

* "Multifunctional nanoparticles – properties and prospects for their use in human medicine," Trends in Biotechnology Vol.26 No.8, 425-433, (2008)

High priority areas include the discovery of ways to measure and quantify the many types of forces between the particles in a wide range of modifying systems. For example, characterization of the nature and complexity of surfaces, both idealized and pragmatic, in concert with the interactions and effects of defects, is a challenge in manufacturing of nanotechnology-enabled products. Effective treatment of nanoscale building blocks and systems needs to be identified and developed to allow their practical usein applications.

3.1.1. Current Advances

Current advances in surfaces, interfaces, and non-bonded interactions of nanomaterials are enabling the use of nanomaterials in medical applications as well as other fields. Exhibit 3.1.1-a illustrates many exciting possibilities and opportunities being explored. For example, optical imaging techniques are used extensively in clinical diagnosis. However, the organic fluorophores in common use are not photostable and have low intensity. The use of fluorescence proteins or bioluminescence systems is also limited because these cannot be optimized in multicolor assays. Nanoparticles are being used today to overcome these problems.[8]

Even small variations in processing of nanoparticles can result in variations of surface chemistry, which may have a significant effect on performance as well as safety. This sensitivity has led to widely variable and conflicting results because it is difficult to make precise measurements of surface chemistry of nanoparticles that have both chemical resolution as well as statistical significance.

As shown in Exhibit 3.1.1-a, there are already several nanoparticle-based systems in use for medical diagnostics, and future possibilities for nanoparticle applications in this field are almost unlimited. The unique properties of nanoparticles enable their use as selective tags or markers for a range of medical targets such as bacteria and individual molecules (e.g., proteins, DNA). Drug delivery is another area where nanoparticles provide many advantages, such as enhancing drug-therapeutic efficiency and pharmacological characteristics. Nanoparticles can improve the solubility of poorly water-soluble drugs, increase drug half-life, enable a more controllable release, and allow the delivery of two or more drugs simultaneously for combination therapy.

3.1.2. Characterization Instruments

Electron microscopies—scanning electron microscopy (SEM) and transmission electron microscopy (TEM), proximal probe microscopies such as scanning tunneling microscopy (STM), atomic force microscopy (AFM), scanning probe microscopes (SPMs), and to a lesser extent near-field scanning optical microscopy (NSOM)—are currently used in nanoscience and nanotechnology research. Line-width measurements are performed by chromatic dispersion (CD), SEMs, CD atomic force microscopes, and optical scatterometry tools. Special tools also exist for measuring pattern distortion and overlay. These tools provide the ability to image the samples and view sample surfaces on the atomic scale as well as with the ability to *manipulate* the particles (atoms, molecules, clusters, etc.) that make up the system.[9]

Over the last two decades, research efforts have sought to provide tools capable of performing measurements on complex, heterogeneous, nanometer-scale systems that are similar to those provided by conventional spectroscopic techniques, but with an increasing level of chemical detail. Advances in proximal probes have demonstrated that STM operation at ultrahigh vacuum and low temperature can achieve spatial resolution to approximately to the atomic level. Microfabricated cantilevers allow AFM to be performed on soft sample surfaces without perturbation. Through intense laser sources and methods for generating small optical apertures chemical contrast mechanisms, dielectric spectroscopy and nonlinear spectroscopy have been demonstrated at length scales well below the diffraction limit of light.

A summary of some of the available technologies for nanomanufacturing is shown in Exhibit 3.1.2-a. Recent advances in TEMs have resulted in resolutions below 0.1 nanometer (nm). These tools are applicable to research and a limited set of manufacturing applications. For SEMs and CD SEMs in manufacturing, the primary problem is image artifacts (charging or scattering of secondary electrons). Although scanning tunneling microscopes have achieved resolution below 0.1 nm, they are not as effective for samples with any topography. Atomic force microscopes and CD atomic force microscopes are limited by the size and variability of tip geometry, which can cause large artifacts.

Exhibit 3.1.2-a. Instrumentation for Nanomanufacturing

Scanning Probe-Based	Beam-Based	Photon-Based
Physical probes that either contact the sample directly or are controlled to be in near contact with the surface	*Particle beams interacting with the sample for determination of the properties of interest*	*Photons of variousenergies ranging from optical to x- and gamma-ray sources*
• Magnetic force microscope	• Scanning transmission electron microscope (STEM)	• Scanning near-field optical microscope
• Magnetic resonance force microscope	• Electron energy-loss spectrometer (EELS)	• Scanning interferometric apertureless microscope
• Chemical AFM	• Scanning electron microscope (SEM)	• X-ray or gamma-ray scattering or diffraction methods
• Dynamic force microscope		
• Cantilever sensors	• Aberration-corrected transmission electron microscope (TEM)	• Photonic force microscope
	• Focused ion beam (FIB)	• Scatterometry

3.1.3. Key Barriers

Specific barriers limiting the understanding and characterization of nanomaterial surfaces, interfaces, and non-bonded interactions identified in the workshop are detailed in Exhibit 3.1.3-a. In summary, these include:

Regulatory, Legal, and Organizational Hurdles. In general, results from past research efforts are not readily available to the industrial and research communities. Many publications exist, but don't have adequate technical details to translate to manufacturing. Development of databases and publications with research outcomes would enable researchers to identify promising areas for further study and also learn from unsuccessful studies. Additionally, cross-cutting research needs are difficult to identify due to a prevailing lack of information-sharing across industrial sectors. Concerns over intellectual property ownership also contribute to this lack of collaboration. A legal framework that reduces R&D partnership risks would help overcome many of the barriers to cross-industry cooperation.

Effective Characterization and Measurement Methods and Tools. Several factors contribute to the inability to characterize nanomaterial chemistries and interactions, particularly the lack of standard protocols and measurement tools. Surface treatments often "mask" the nanoparticle, thereby modifying the raw particle's attributes and making it difficult to characterize. Better measurement techniques that ensure proper isolation of the nanoparticles would help to determine the effects surface treatments.

Exhibit 3.1.3-a. Key Barriers For Surface/Interfaces and Non-Bonded Interactions of Nanomaterials

Barriers	Needs	Priority
Regulatory/Legal/ Organizational	Collaboration across disciplines	•••••
	Databases or other publications that capture what works and what doesn't	•••
	Legal framework for companies to reduce risk	••
	Database for requirements and needs	
	Published results for poorly characterized materials	
Characterization and Measurement	Standard protocols/assays (instrumentation)	•••••
	Nanoparticles – not a "masked", complex system – for dispersal	•••
	• Blend particles without distorting attributes	
	• Maintain functionality	
	• Online or surrogate monitoring method and validation	
	Adequate measurement tools for chemists (for distribution, etc.)	••
	Field techniques to measure quality or characteristics	•
	• Practical methods, to lab and back	
	Better methods and standards for protocols	
	Statistical significance of nanotechnologymeasurements	
	• Atomic resolution	
	• Cost effective tests	
	Metrics to confirm that raw nanoparticles are being studied	
Modeling	Modeling exchange, sharing of information, and framework of validated experimental structure-property data to develop and validate models	•••••••
	• Identify when nanotechnologytakes over	
	• 2D versus 3D	
	Transferability – uncertainty of models available	•
	Control or shorten half life (for functionality)	
	Predict functionality over time	
	• Model nanoparticle interaction with cells	
	Understand interactions between various systems (e.g., food, other systems)	
EH&S	Insufficient studies, basic science (is toxicity a result of surface interactions?)	•••••
	• Aspect ratio, diameter, functionality	
	• Biocompatibility	
	Understand interaction and synergy with biological systems	••••
Performance Properties	Basic science focused on market pull	•••••
	Scalable process for complex, large areas	•••
	Maintain functionality over the long term	•••
	Determine what particles are practical and safe to use now	•
	Reliable products	•
	Cost models	
	On-line measurement and characterization	
	Control stability over full life cycle	
	Ensure particles have desired/required characteristics and properties	

Modeling. Identifying ways to enhance modeling techniques is inhibited by a lack of coordinated R&D needs, information exchange, and systematic experimental structure-property data to develop and validate models. Modeling improvements would enable researchers to predict nanomaterial functionality and improve product reliability at lower cost and in less time. Other technical enhancements would address the nanomaterial properties for testing, such as controlling or shortening their half-life for optimal or shorter testing periods.

Environmental, Health, and Safety Issues (EH&S). Current research has not been able to establish a link between nanomaterial surface characteristics and toxicity. Further studies are needed to determine the influence of surface interaction on toxicity.

Performance Properties. Several challenges contribute to the inability to link nanomaterial surface characteristics to performance properties. Priority areas needing additional study include: preventing the time-dependent loss of functionality of nanomaterials and the development of processes that enable scaled-up production and testing. Other high priority areas include the inability to quickly characterize nanomaterials and to identify which nanomaterials are practical, safe, or reliable to use in manufacturing.

Process Control. Small variations in process conditions can result in nanomaterials with variable surface chemistry and structure that significantly impact performance and toxicological properties. When manufacturing products contain nanomaterials, these variations in surface chemistry can lead to nanomaterial aggregation and nonhomogeneous distribution that result in poor performance and product reliability. In composites, for example, uneven distribution of nanofillers can create weak areas that become points of failure. Currently it is a challenge to measure the surface characteristics and distribution of nanomaterials in a manufacturing stream or composite with the resolution and statistical significance required for reliable manufacture and performance that is essential for widespread commercialization.

3.1.4. Priority Solutions

Cross-industry solutions to address the barriers were discussedand alist of topics identified, as follows:

- Developing new measurement science
- Developingnew measurement tools
- Fostering collaboration across disciplines
- Modeling exchange and information sharing
- Measuring toxicity as a result of surface interaction
- Maintaining long-term functionality (for reliable products)
- Aligning research with market pull
- Informationand property databases

Exhibit 3.1.4-a. Cross-Industry Solutions to Accelerate Progress in Surface/Interfaces and Non-bonded Interactions of Nanomaterials

Topic	Solution	Priority
New Measurement Science		
	• For both measurement techniques and predictive models, develop an inventory, identify gaps, and drive solution development	••••••
	• Identify existing tools andestablish range of applicability	
	• Develop infrastructure framework of systematic experimental structure-property data to develop and validate predictive models	
	• Organize and maintain to identify gaps, range of size, classes of materials and surface interactions, accuracy and precision, range of physical properties	
	• Keep track of new developments in progress	•
	• Focus on problems with market pull and link to fundamental science questions to coordinate activity and drive science forward	
	• Learn/define what to measure, how to measure –from impacts to performance	
	• Test particles and precursors, ensure proper purification	
New Measurement Tools		
	• Identify surrogate measurement tools for online measurement (manufacturing) that can relate to research tools (laboratory)	••••
	• Standard protocols, assays (instrumentation) –field techniques to measure quality	
	• Property databases and information tools	
Fostering Collaboration Across Disciplines		
	• Integrate industry needs and academic funding – establish a mechanism to translate commercial needs into fundamental science	••••••
	• Develop organization of labs that want to test, develop standards for nanomaterials characterization	•••
	• Standard reference simulations for new users	
	• Known systems and validation for tools, limits and range of applicability	
	• Expand mechanisms for industry/Federal research, definition of important problems, co-funding	••
	• Sharing best practices for university/industry collaborations/IP, etc.	

Exhibit 3.1.4-a. (Continued)

Topic	Solution	Priority
Modeling Exchange and Sharing of Information		
	• Integrate modeling and experiments – increase collaboration	• • • •
	• Initiate model verification/validation for nanomaterial surface interactions (e.g., nanoparticle/cell membrane interactions; nanoparticle-solvent interactions to validate pair-potentials to quantify noncovalent interactions in atomistic simulations)	• • •
	• Need blind modeling competition, mapping space for a well-known nanocomposite systems	• • •
	• Create programs that will generate basic properties needed for models	
Toxicity and Surface Interaction		

	• List existing toxicity interactions/assays
	• List existing measurement tools efforts
	• Inventoryexisting modeling efforts before going forward
	  • Pose issue/problem, call for ideas/solutions – establish an interactive working group process to achieve this
	  • Protect competitive aspects
	  • Model known systems to validate modeling;understand limits and range of applicability
	  • Utilize standard reference simulation to train new users
		• • • •

Maintaining Functionality/Reliability Long Term		
	• Correlate fundamental nanoparticle surface understanding to performance properties • Understand surface and interfacial properties to ensure, control, and measure dispersion – initiallyand in the long term	• • • • • •
Market Pull		
	• Develop reference systems relating to market needs • Need mechanism to facilitate/encourage university/academic research and strengthen collaboration on market-relevant materials • Identify nanotechnologyattributes thatclearly relate to specific market space	• • • • • •
General		
	• Fund basic research on surface/interface properties and relationships with toxicity	• • •
	• After problems are defined, develop infrastructure to carry out research	•
	• Fund fundamental research on surface and interfaces of nanoparticles – hydrophophic, chemical, Van der Waals, etc. • Fund basic research to develop new measurement tools	

EXHIBIT 3.1.4-B. PRIORITY CROSS-INDUSTRY SOLUTIONS SURFACES, INTERFACES, AND NON-BONDED INTERACTIONS OF NANOMATERIALS

- Predict and control toxicity of nanoparticles
- Develop reference systems relating to market needs
- Establish a network of labs that test and develop standards for nanomaterial characterization
- Research on surface interfaceproperties relevant to performance
- Collect list of existing models, measurement tools, and biological interaction assays
- Identify or develop surrogate measurement tools for on-line measurements that relate to research tools
- Integrate industry needs and academic funding—mechanism and framework to translate commercial needs into fundamental science questions
- Correlate fundamental nanoparticle surface understanding to performance priorities
- Improved computational models for nanomaterial surface interactions
- Measurement techniques and predictive models interaction: identify gaps and devise solution development
- Increase collaboration in modeling and experimentation

Each topic was explored and the importance of possible solutions ranked (see Exhibit 3.1.4-a). While some of the solutions focus on a specific barrier, others can apply across multiple barriers (e.g., development of new measurement tools can improve modeling capabilities and enhance performance properties).

From the solutions identified in Exhibit 3.1.4-a, a set of priority solutions was selected and grouped as shown in Exhibit 3.1.4-b. Mini-roadmaps for each solution are shown graphically on the pages that follow.

3.2. Nanotechnology-Enabled Composites and Matrices

One of the keys to unlocking the benefits of nanotechnology is building the nanomanufacturing science and technology base to the point where nanomaterials exhibiting unique nanoscale properties can (1) be designed based on existing and emerging science; (2) routinely be placed into components or systems; (3) retain and combine their unique nanoscale properties in a matrix of other materials; and (4) result in superior and controllable composite performance. In developing the needed nanomanufacturing technologies, greater commercial influence and awareness serves to guide research into the highest priority and most productive areas.

SURFACES, INTERFACES, AND NON-BONDED INTERACTIONS OF NANOMATERIALS

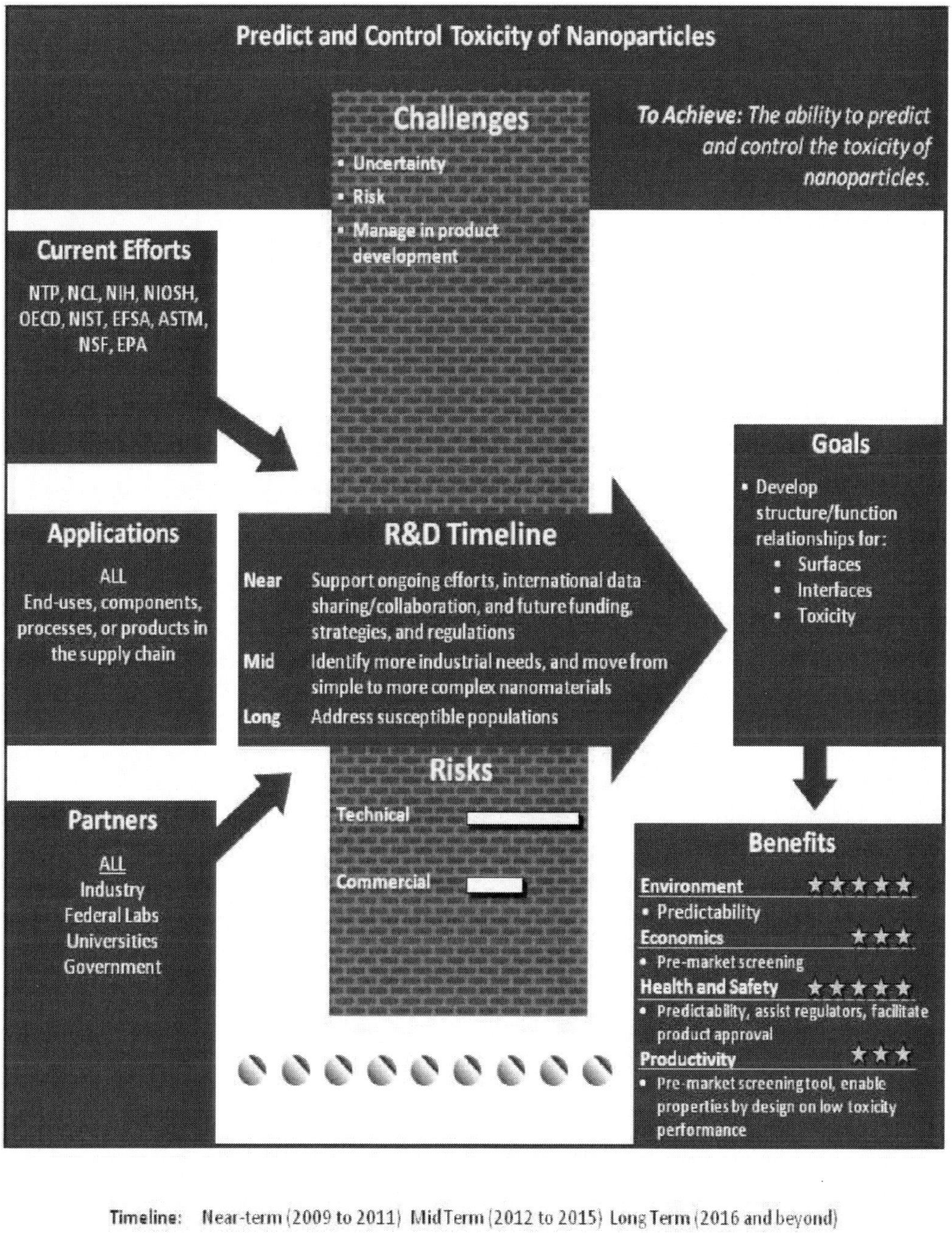

Priority Solution: Predict and Control Toxicity of Nanoparticles

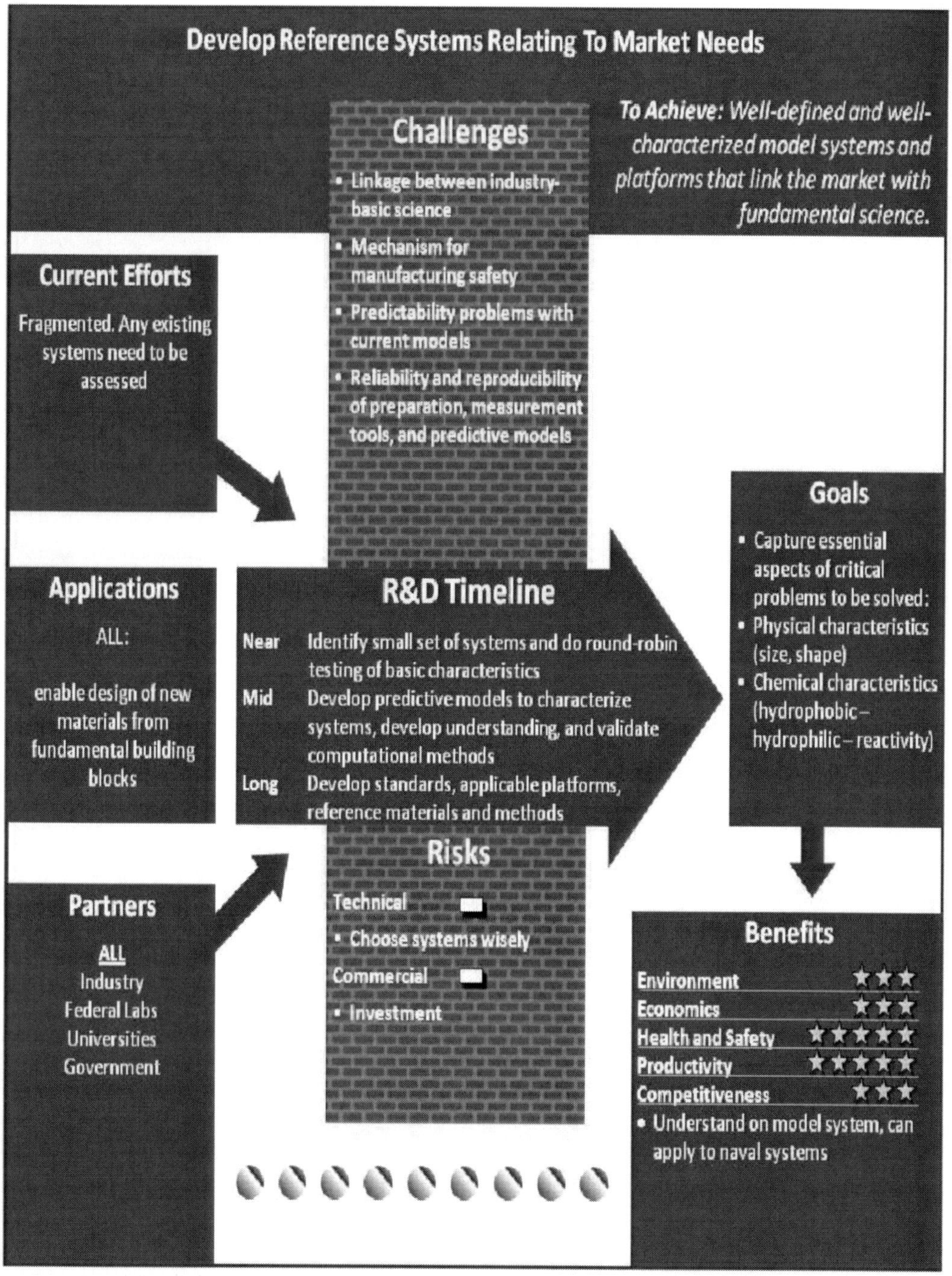

Timeline: Near-term (2009 to 2011) Mid Term (2012 to 2015) Long Term (2016 and beyond)

Priority Solution: Develop Reference Systems Relating To Market Needs

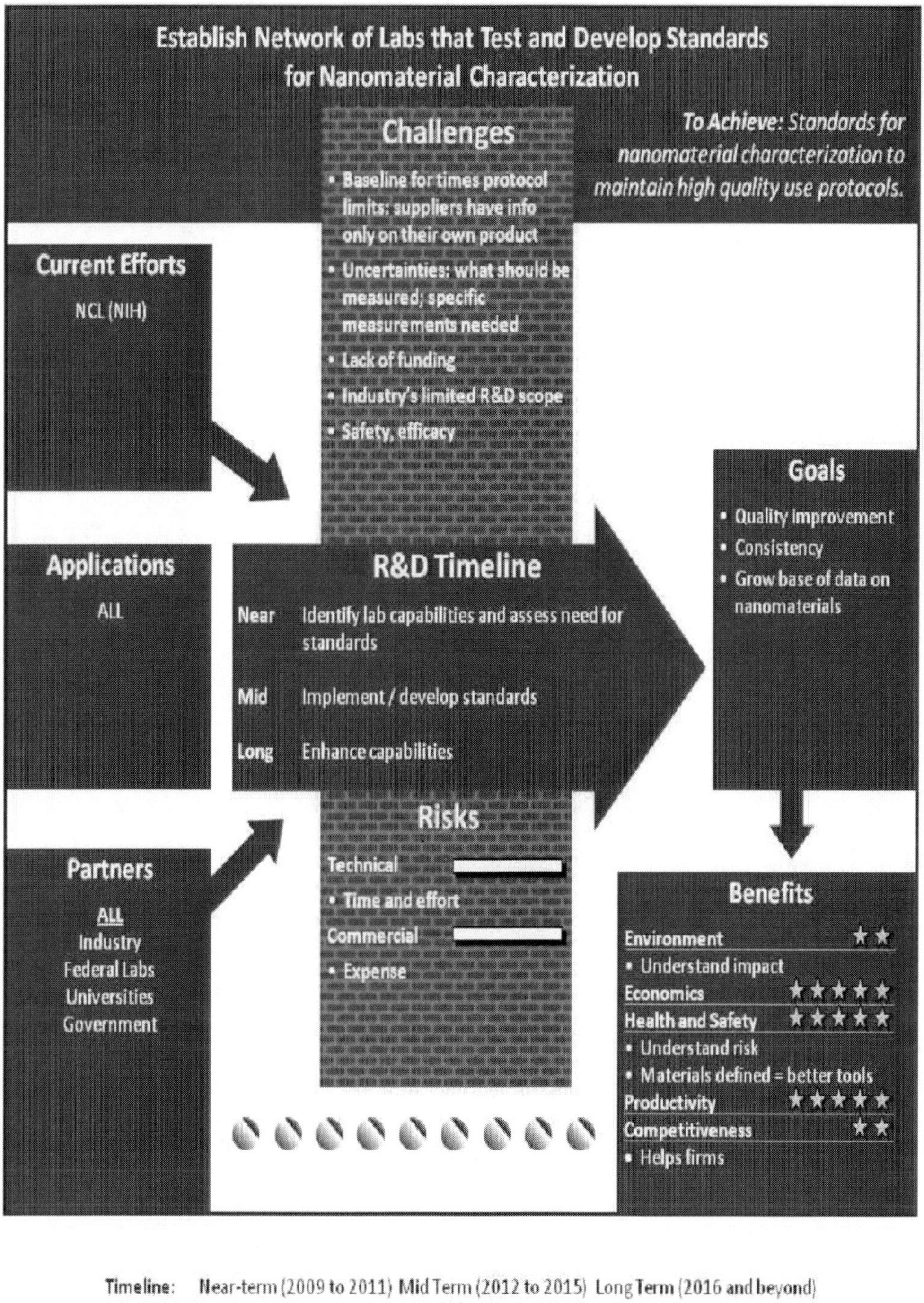

Priority Solution: Establish Network of Labs that Test and Develop Standards for Nanomaterial Characterization

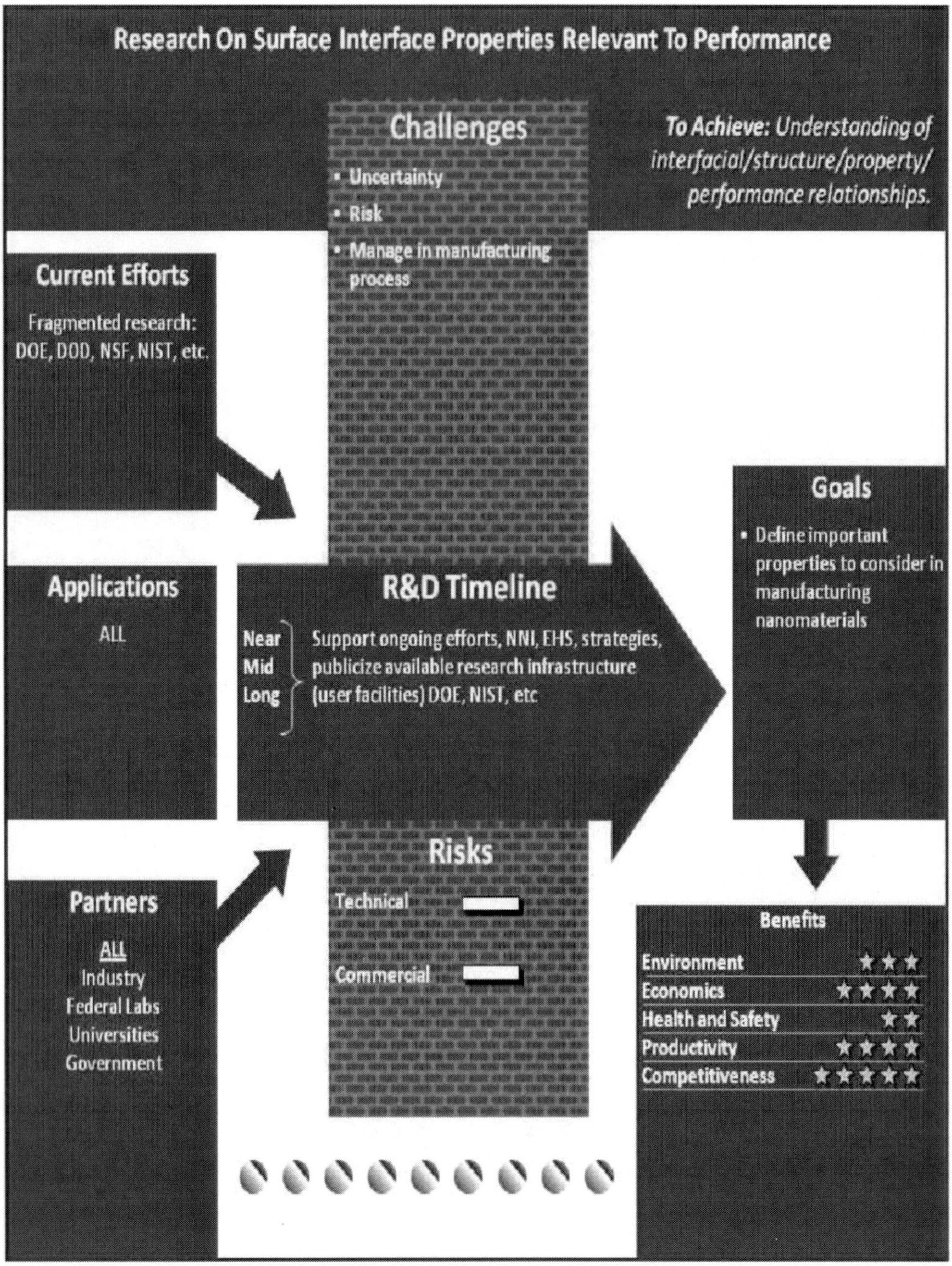

Timeline: Near-term (2009 to 2011) MidTerm (2012 to 2015) LongTerm (2016 and beyond)

Priority Solution: Research On Surface Interface Properties Relevant To Performance

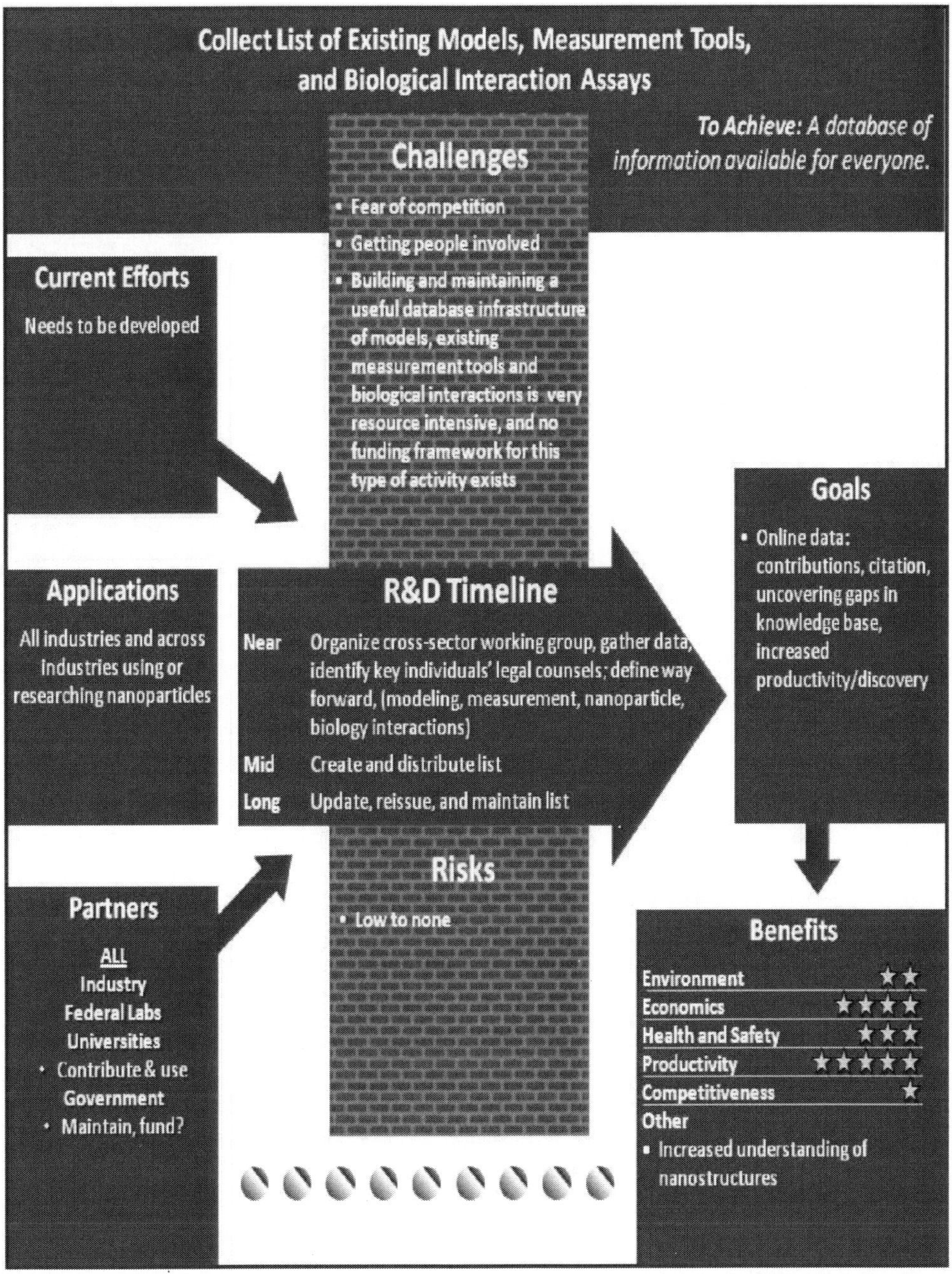

Timeline: Near-term (2009 to 2011) Mid Term (2012 to 2015) Long Term (2016 and beyond)

Priority Solution: Collect List of Existing Models, Measurement Tools, and Biological Interaction Assays

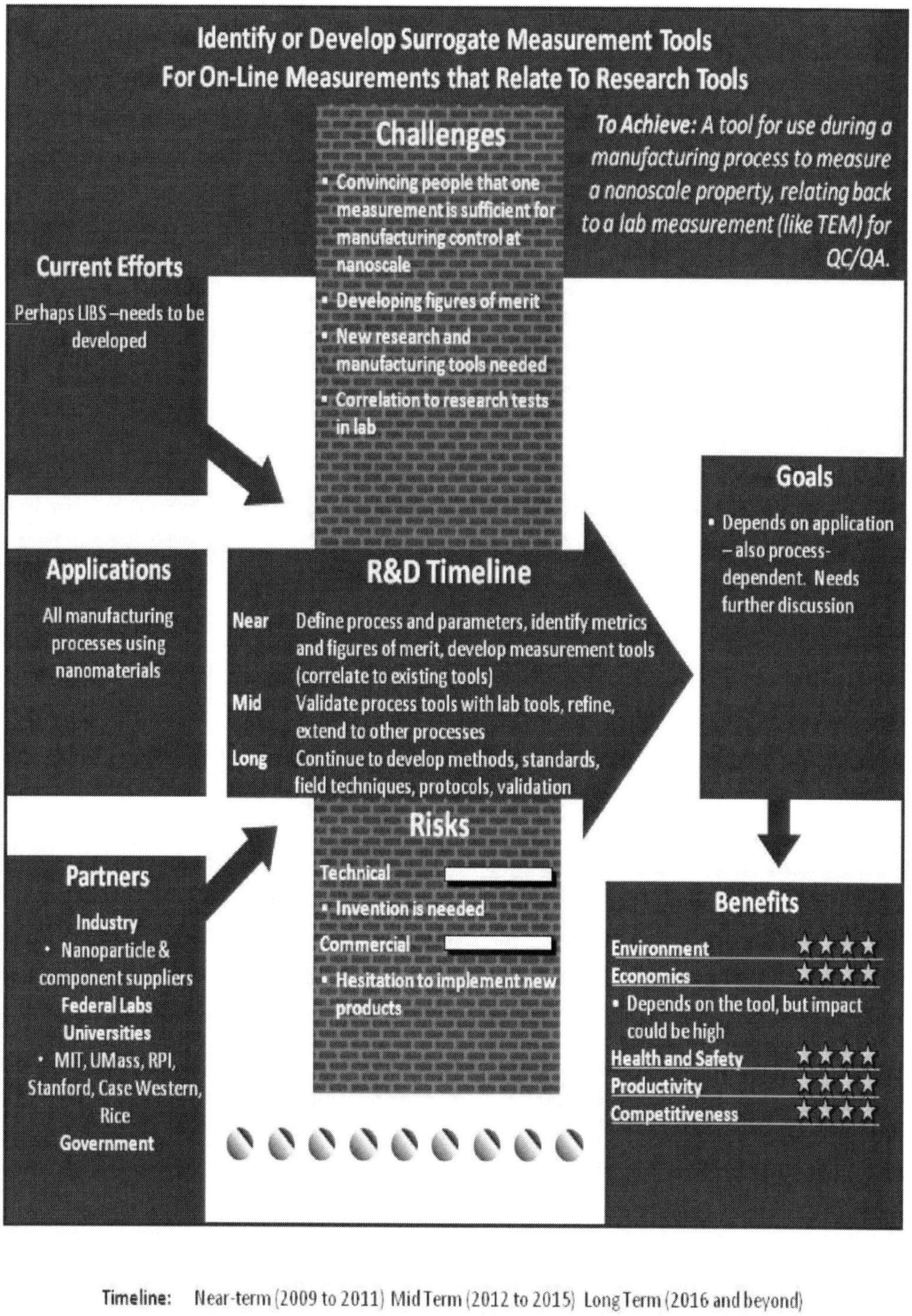

Priority Solution: Identify or Develop Surrogate Measurement Tools For On-Line Measurements that Relate To Research Tools

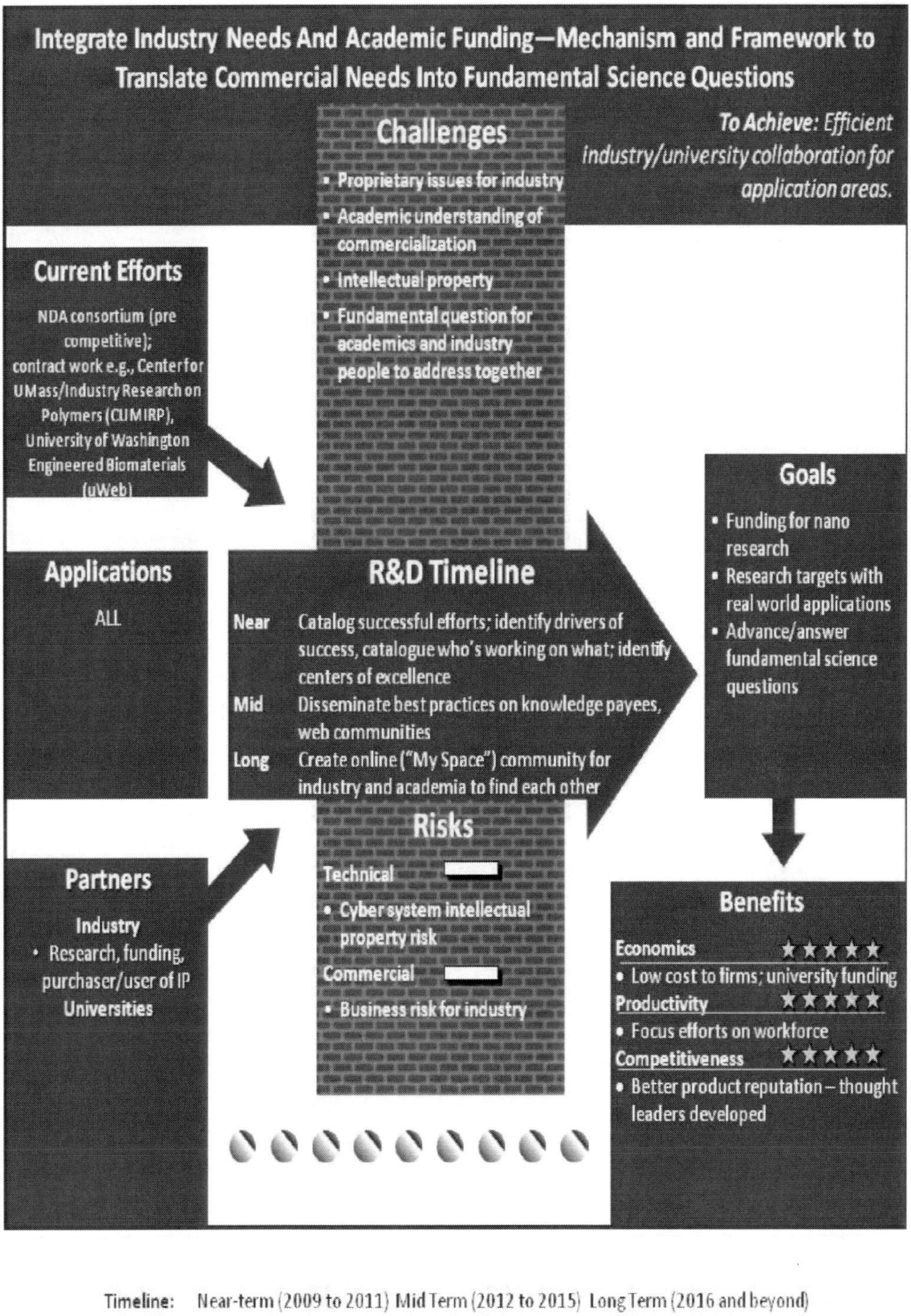

Priority Solution: Integrate Industry Needs And Academic Funding—Mechanism and Framework to Translate Commercial Needs Into Fundamental Science Questions

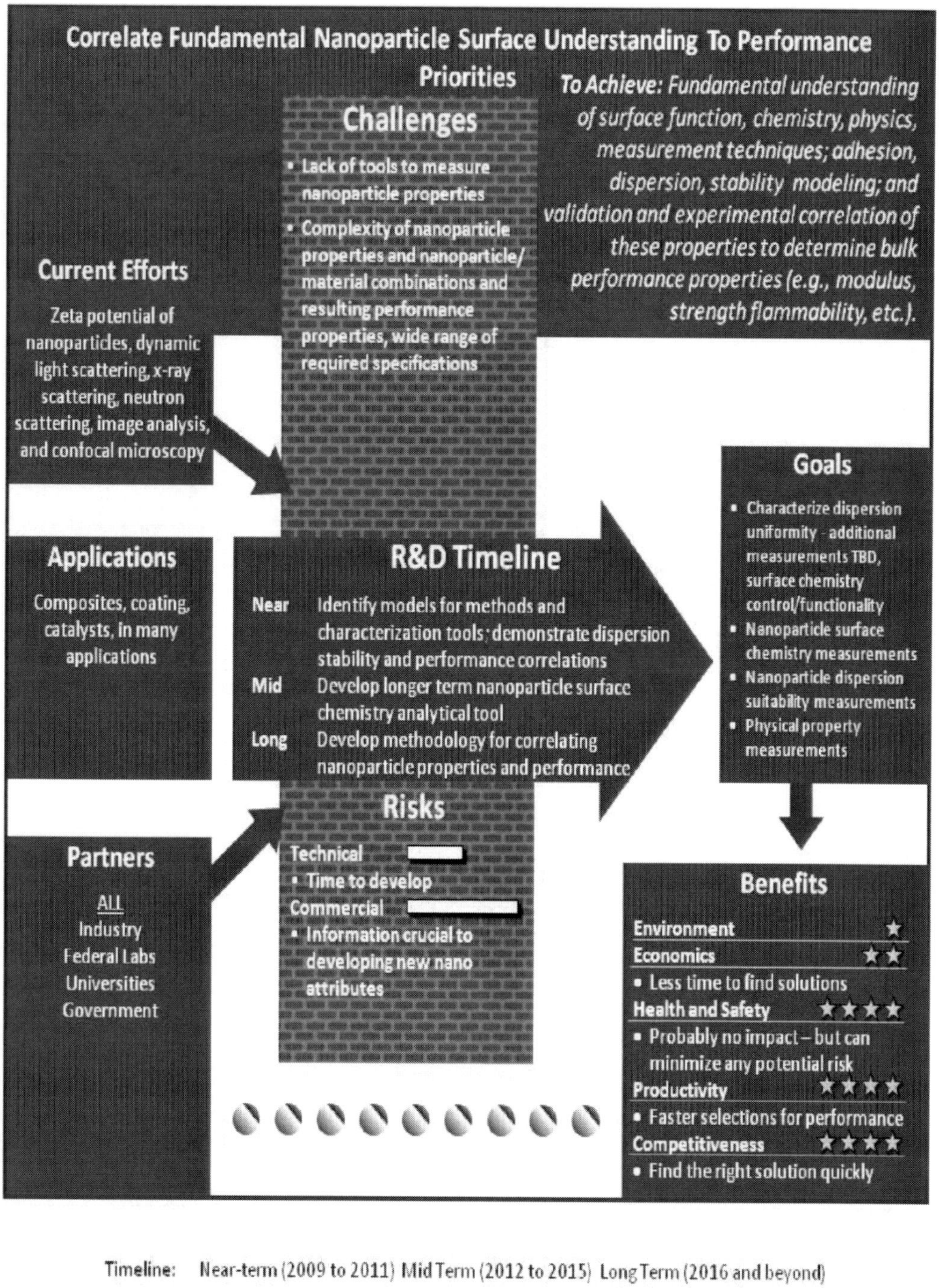

Priority Solution: Correlate Fundamental Nanoparticle Surface Understanding To Performance Priorities

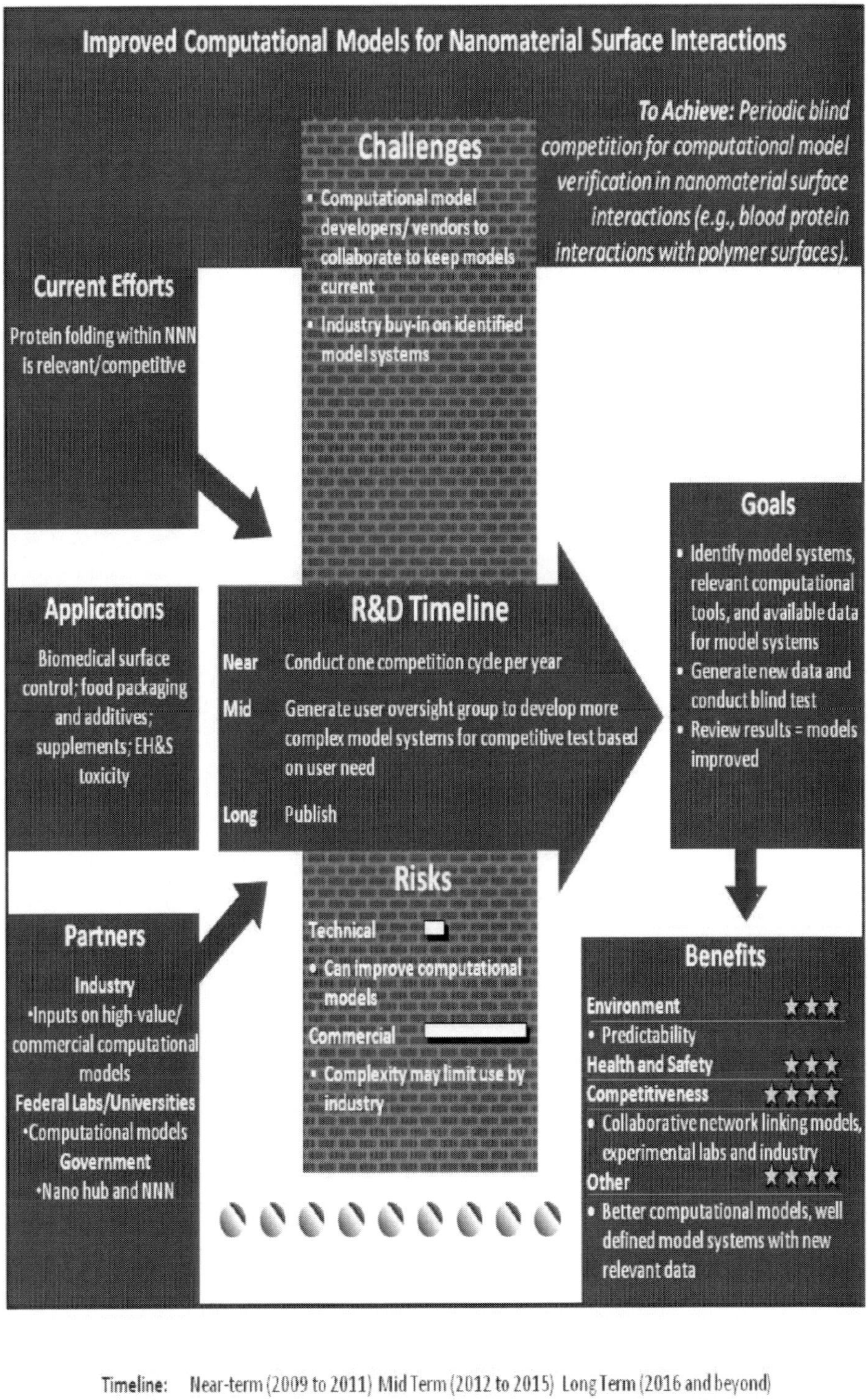

Priority Solution: Improved Computational Models for Nanomaterial Surface Interactions

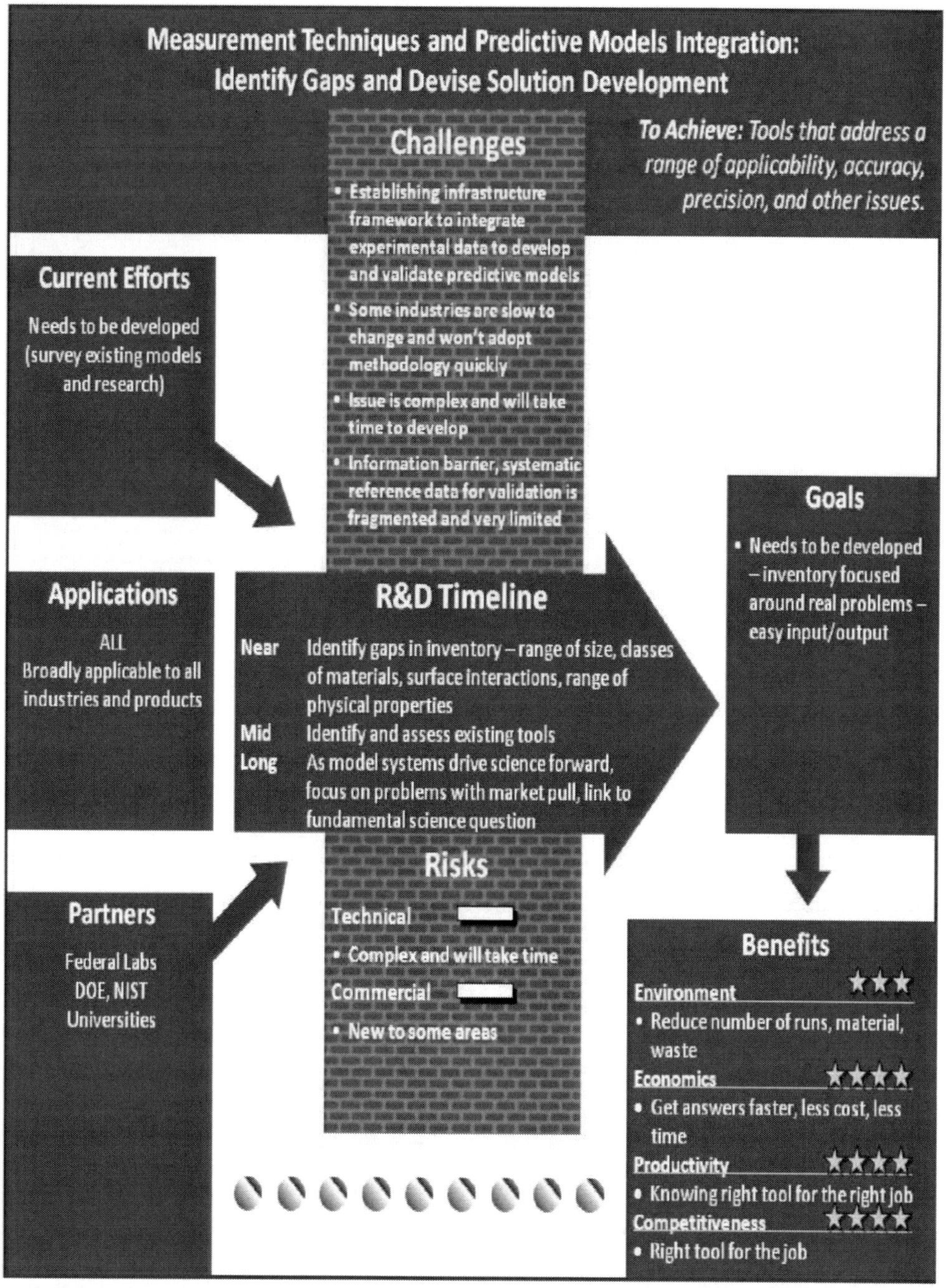

Timeline: Near-term (2009 to 2011) Mid Term (2012 to 2015) Long Term (2016 and beyond)

Priority Solution: Measurement Techniques and Predictive Models Integration: Identify Gaps and Devise Solution Development

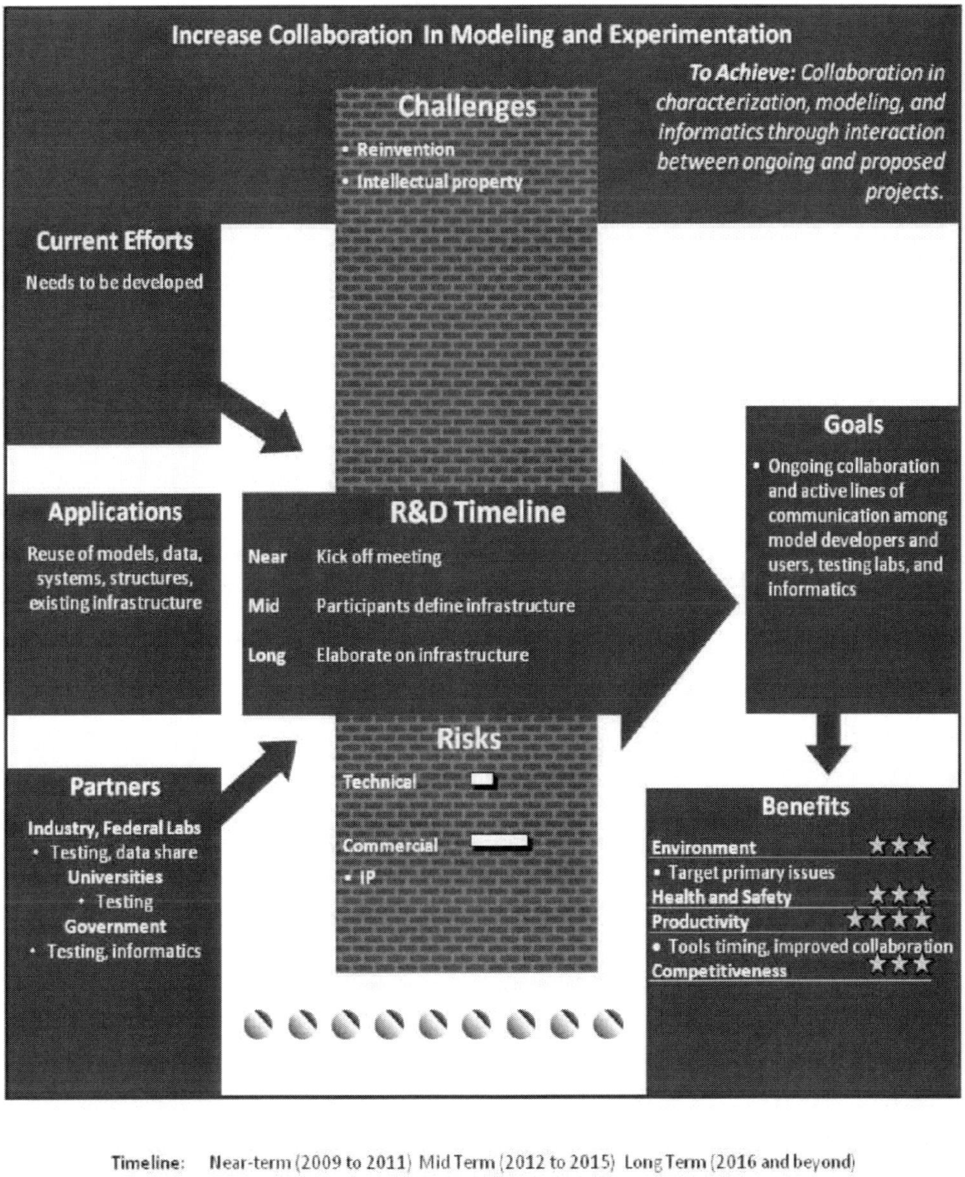

Priority Solution: Increase Collaboration In Modeling and Experimentation

Nanomaterials can boost energy efficiency in a wide range of applications by offering such properties as light weight, high mechanical strength, unique color, electrical properties, and high reliability in extreme environments. Applications could be as diverse as biological implant materials, electronic packages, and automotive or aircraft components. Although some properties will be common among applications, others will need to be quite different. For example, an electronic package polymer composite must be electrically insulating, while an aircraft component may need to be electrically conductive to dissipate charges from lightning strikes.

The addition of small amounts of nanoparticles to polymers can enable new properties in the composite material, but results are highly dependent on the surface treatment of the nanoparticles and processing used. It will be valuable to determine whether nanomaterials could be integrated into nanocomposites to enable multiple desirable properties required for a given application.

While industry is seeking materials with unique properties to address difficult challenges, no "rule of mixtures" exists as a guide for mixing multiple nanomaterials to obtain all required properties in a composite structure. Nanomaterials often have unique properties that could enable composite materials with multiple unique properties simultaneously, yet it is often challenging to achieve these properties in large-scale nanocomposite materials. Furthermore, nanomaterials should provide desirable properties that cannot be achieved through use of conventional chemicals and materials.

3.2.1. Current Advances

Progress in nanocomposites is varied and covers many industries. Nanocomposites can be made with a variety of enhanced physical, thermal, and other unique properties. They have properties that are superior to conventional microscale composites and can be synthesized using simple and inexpensive techniques. To assess the potential value of nanocomposites, developers must determine which nanomaterials can be effectively integrated into nanocomposites and what new or improved properties this enables. This process entails determining the effectiveness of dispersion of the nanoparticles in the matrix and how this affects the structure of the polymer to enable optimization of the desired property. Once the basic models of this process are developed, it must be determined how the mixing of multiple nanomaterials in a polymer affects the resulting structure and properties of the nanocomposite. One nanomaterial may be required to improve the mechanical property, and another may be required to change the electrical properties. However, the addition of the electrical material may also change the mechanical properties of the nanocomposite through interactions with the polymer and nanoparticles.

SESSION SCOPE

- Identify science foundation needed to enable nanocomposite materials with unique properties
- Develop knowledge and models to enable design of nanocomposites with multiple properties optimized simultaneously within these topical areas
- Discuss and refine these needs and arrive at a prioritized list of fundamental science gaps with respect to the needs
- Develop knowledge to enable scale-up of nanomanufacturing (may be a competitive capability)
- Define the challenges and barriers for filling these gaps and propose solutions

In general, two idealized polymer-layered nanocomposite structures are possible: intercalated and exfoliated. The greatest property enhancements are generally observed for exfoliated nanocomposites. These consist of individual nanometer filler layers suspended in a polymer matrix. In contrast, intercalated hybrids consist of well-ordered multilayers with alternating polymer/nanometer filler layers with a repeat distance of a few nanometers. In reality many systems fall short of the idealized exfoliated morphology.

Existing Materials. A few nanocomposites have already reached the marketplace, while others are on the verge. The global nanocomposites market is projected to reach 989 million pounds by the end of the decade.[10] The United States and Europe dominate the global nanocomposites market, with a collective share of over 80% of the volume sales for 2008.

Nanocomposites influence many industries, such as computers, electronics, plastics, coatings, magnets, water remediation, energy, and medical equipment. For example, a nanoscale memristor switch has been developed using nanocomposites. The memristor—short for memory resistor—could make it possible to develop far more energy-efficient computing systems with memories that retain information even after the power is off, so there's no wait for the system to boot up after turning the computer on. It may even be possible to create systems with some of the pattern-matching abilities of the human brain.[11]

The uses for nanocomposites are highly diverse, as the following examples illustrate. Using a nanocomposite ZnO/polymer nanocomposite, Applied Nanomat, Inc. (ANI) scientists have invented the world's first nanoscale electric generator. This ground-breaking invention is a revolution that allows us to harvest mechanical energy from various microenergy sources in the environment and human body, then convert them into electric energy.[12]

Progress has also been made in solar power based on nanostructures and nanocomposites. For example, dye-sensitized solar cells using micro/nanocomposite TiO porous films resulting in cells with enhanced light collection, have been fabricated. This technique opens an alternative way for manufacturing solar cells on an industrial scale. Other solar developments include development and manufacture of revolutionary nanostructured ultra thin film solar (photovoltaic-PV) products that increase the total PV surface area by an incredible 6-12 times over other thin film products on the market today.

Graphene nano-platelets have been developed as additives for advanced composites, as a substrate for advanced electrical or electronic applications, as the conductive component in specialty coatings or adhesives, and as a component of e-inks or printable electronic circuits. Low-cost processes have been engineered that provide the ability to tailor multi-functional materials that take advantage of exceptionally large surface area to mass ratios of active metals and metal oxides for applications such as catalysis, lithium-ion battery electrodes, supercapacitors, solar cells, and energy storage.[13]

Nanocomposite permanent magnet materials are a new type of permanent magnet material consisting of magnetically hard and soft grains, both in nanometer size.These materials have a high potential to be developed into high-performance permanent magnets with very high energy production.[14]

Instrumentation/Equipment. There are many current efforts underway in manufacturing, testing, and simulation technology for nanocomposites. A few are outlined below.

For *manufacturing of composites*, plasma coupling technology provides the capability forming carbon nanotube/fiber systems that are highly flexible and reproducible (singlewall/multiwall), and of nanofiber deposition. Capacitive coupling requires a high capacitance between the electrode and the plasma (large amplitude radio frequency (RF) voltages). The inductive coupling requires a high inductance between a coil and the plasma (large RF currents).[15]

Nonradiative energy transfer (incoherent) relies on *long-range electrostatic* Coulomb interactions that can potentially enable highly parallel, defect tolerant, and easily scalable communicating structures. On the other hand, strong electrostatic "plasmonic" interactions can be used for controlling energy flows in metal nanoassemblies. The ability to predictably control interfacial electrostatic interactions can lead to such important applications as electrically pumped tunable light emitters, on-chip plasmonic circuitry, artificial photosynthesis, low-cost photovoltaics, Terahertz detectors, and bio/chemical sensors.[16]

Molecular Vapor Deposition (MVD®) technology enables the growth of ultra-thin films with a wide range of functionalities on a broad spectrum of substrates. MVD is a breakthrough proprietary nanotechnology technique that allows for room temperature vapor deposition of several types of organic and organometallic molecules. This innovative technique has facilitated many new emerging applications with small feature sizes that have been difficult or impossible using traditional liquid synthetic techniques.[17]

For *testing of composites*, the mechanical response of nanoscale materials and structures has important implications, from understanding biological recognition and development of lightweight structural materials, to exploration of new concepts for switches and chemical sensors. While instrumentation such as nanoindentation exists to measure mechanical properties of nanostructured films on substrates, the development of techniques to reliably measure mechanical response of structures with nanometer-scale architectural dimensions such as nanowires, nanotubes, nanodots, nanofilms, nanopillars, nanoporous, etc, remains a challenging problemin nanomechanics.

New Materials and R&D Advances. Ongoing research generally focuses on state-of-the-art developments and the possible environmental impact of nanocomposites. For example, a new class of optical materials known as left-handed (or negative index) metamaterials that are not found in nature are being explored. These structures can focus light down to dimensions that are far beyond the classical diffraction limit.[18]

In other areas, research is ongoing to leverage attributes of carbon nanotube-based conductive coatings to solve real world problems for the security, medical, and performance coatings markets. This includes materials that can significantly improve the performance of commercial and military lithium ion batteries through the development of new salts.[19]

New Y-Carbon technology allows production of nanoporous carbon of uniform and controlled pore size. By varying different precursor and processing parameters, pore size can be tuned to sub-nanometer (about ten billionth of a meter) precision, something that is unattainable with the conventional route of carbon synthesis.[20]

Research is also ongoing to evaluate the impacts of nanomaterials in general on the environment, health, and safety. There is a significant body of work in this area, and numerous efforts are in progress to identify and define an effective approach for assessingthe potential impacts of these new materials.[21]

3.2.2. Key Barriers

Specific barriers limiting the development of nanotechology-enabled composites and matrices identified in the workshop are detailed in Exhibit 3.2.2-a. In summary, these include:

Achieving nanotechnology-enabled composites/matrices with multi-functional properties. Industry requires better understanding and control over the functionality of nanomaterials in matrices before it can cost-effectively use their unique combinations of properties (e.g., strength, optical, magnetic, and electrical) to improve final products. Industry wants to better understand how performance properties vary as a function of size and surface area. Advances in this area could potentially enable breakthrough applications such as load-bearing windows, lightweight drilling rigs, super capacitors (high charge density batteries), and drug delivery.

Modeling properties across dimension scales. A key enabling technology would be to have predictive modeling capability that accurately relates material properties (mechanical, electrical, thermal, etc.) and the combinations of properties from the atomic- to the macro-scale. What is needed is a series of modeling techniques, each experimentally verified, that bridge across the atomic to macroscopic length scales, such as, but not limited to, quantum mechanics, molecular dynamics, meso-dynamics, phase fields, and finite element analysis. High-quality models would make it possible to accurately predict composite properties from the nanotechnology additives. Success in this area would mean fewer materials to be tested and simulations that could accurately predict unexpected properties.

Measuring the mixing and dispersion of nanoparticles. Thorough dispersion of nanoparticles within a matrix or composite is essential to deliver the intended properties of a nanomaterial, yet current tools are inadequate to provide the required level of quality assurance for such dispersion. A key difficulty is the inability to assure true dispersion and the absence of agglomeration of such tiny particles throughout an entire matrix or composite. A tool is needed to provide an accurate and realistic characterization of the dispersion of nanoparticles.

Scaling up for large-scale production/manufacture. Techniques for precision control of the scale-up process are needed to reliably preserve nanomaterial properties in large-scale manufacturing. Batch, continuous, and campaign industrial manufacturing applications require that large amounts of nanomaterials be completely and predictably consistent in makeup, properties, and quality over extended periods of time. Process control measurements and processes designed to take advantage of nanoscale interaction are needed, as well as processes that direct nanotechnologyself assembly.

Effectively assessing and communicating EH&S risk. Industrial manufacturers are necessarily concerned about exposure to liability. They need to know what the risks of exposure are to human health and safety and to the environment. They need a way to reliably measure the presence of both manmade and preexisting nanoparticles in the work environment and in end-use environments. Supporting analyses may require methods for artificially aging or accelerating effects of time and weather on nanoparticles and nanomaterials.

Bonding nanofibers for high strength. Enhanced strength is a desirable property often associated with nanomaterials. The ability of nanotechnology to deliver on this promise requires a better understanding of the best ways to connect the nanofibers. Controlled self-assembly techniques that deliver high strength would enable multiple manufacturing applications, potentially including enhanced electrical conduction via nanotubes.

Exhibit 3.2.2-a. Key Barriers to Use of Nanotechnology-Enabled Composites and Matrices Across Industry

Barriers	Needs	Priority
Value	Valuable early prototypes • Making even one useful nanotechnology-enabled material (e.g., aerospace) • Results to justify continued or expanded Federalfunding • Sufficient information to enable decisions by corporate managers	●●●●●●●
	Valid value propositions • Show value in use • Define market needs • Acceptable changes to customer's production process • Define cost structure • Overcome cost concerns about engineered nanomaterials	●●●●●●●
	Certification/approval authorizing agents unaware of differences between composites and metals • Stringent government regulation will stifle/inhibit research, development, and deployment (RD&D)	●●●●
	Under-developed supply chains • Gap between nanomaterials producers and application companies Finding applications for lab results	●
Multi-Scale Modeling	How to relate properties from the nano- to the macro-scale	●●●●●●●
	• Finite element analysis to define relationship of structure to molecular properties • Deduce composite properties from nanotechnology additives Predict interactions between filler and matrix Models of nanomaterial effects and usage	●●●●●
Characterization	Property barriers: Inability to • Determine solubility in polymers • Measure leachants and extractables • Measure permeability with small sample sizes	●●
	Achieve/prove consistent synthesis Fundamental understanding of phase separation at nanoscale • Interaction of nanoparticles with polymer phases not understood Multi-functional instrumentation	●
Measurement	Accurate measurement of mixing or dispersion	●●●●●●●
	Ability to measure time-dependent properties and capture effects of aging • Appropriately accelerate conditions (aging) for all elements in nanocomposite • Measure dynamic heterogeneity in nanocomposite	●●

Exhibit 3.2.2-a. (Continued)

Barriers	Needs	Priority
	• Better understanding of properties of semi-crystalline nanocomposites	
	Need database of standardized nanomaterial properties	•
Performance Properties	Ability to build composites with multi-functionality	•••••••
	• Multi-functional properties	••••••
	• Preserving functionality of nanomaterials (e.g., strength, optical, magnetic, etc.) in matrices	
	• Improvement to final product	
	• Control of multiple properties	
	Better understanding and control over electrical and optical properties	••
	Better understanding of performance properties as a function of size and surface area	••
Processing	Scale-up for industrial use	•••••••
	• Controlled technology change needed in existing process, formulator	••••
	• Large-scale manufacturing infrastructure	
	• Process changes/impacts	•••••••
	Challenges in bonding of nanofibers	
	Poor flow of nanoparticles at large scale	
	Ability to control and use in-situ formation	
	Ability to control meso-structure (porosity, homogeneity)	
EH&S	Mechanism for communicating EH&S risk (real or perceived) along the supply chain	•••••••
	Better information on sustainability	•
	• Determine options for proper disposal/ reuse/recycling	
	Reliable, available methods to measure or detect engineered nanoparticles in the workplace or environment	
	Identification of best models/assays for preclinical and toxicological assessment of nanomaterials	

Establishing valid value proposition. Manufacturers require a clear and quantifiable demonstration of the value added by incorporating nanomaterials into their products or processes. In most cases, use of a nanomaterial must be shown to either deliver valuable added functionality at the same or only slightly higher cost than the current process or must deliver the same functionality at a lower cost. While the process for determining value proposition is well established, it has not been applied to nanomaterials. Extensions of the value proposition as it relates to nanomaterials include quality assurance, materials certification, EH&S issues, cost, final properties, and the properties as they affect processing (e.g., nanoparticle loading level in a matrix affects cost, viscosity, flow rate, and processing time). One value-added application could help sell industry on nanomaterials and get them widely invested in the effort.

The results in Exhibits 3.2.2-a and 3.2.3-a indicate that a combination of multi-scale modeling and experimental characterization techniques and informatics are needed to enable the capability to produce nanocomposites with properties-by-design criteria. As this capability is developed, it will be important to assess the potential degradation of nanocomposite properties throughout the life cycle and characterize the interactions of nanocomposites with environmental conditions, such as moisture, temperature, and mechanical stress.

Exhibit 3.2.3-a. Cross-Industry Solutions to Accelerate Progress in Nanotechnology-Enabled Composites and Matrices

Topic	Solution	Priority
Informatics		
	• Develop searchable databasefor nanomaterials and their properties	••••••••
	• Develop nanoinformatics database on current understanding of interfaces at nanoscale	
	• Enable cross-field/industry knowledge-sharing and idea fertilization, focused on science/engineering organizations	
	• Extend Toxnet[22]database with added nanomaterial searching method	
	• Communicate "failures" to research community	••••••••
	• Informatics (broken up by materials class)	
	• Launch major effort (like moon shot)	•••
	• Develop better IP technology transfer models	
	• Improve technology transfer: communication, dissemination, matching	
	• Develop central "nanomanufacturing hub"	
	• Gain knowledge and reduce number of networks for nanotechnology	
	• Perform knowledge gap analyses and focus research on criteria gaps	
Communication		
	• Create better academic/government lab/industrial communication structure	•••••
	• Improve industry coordination in communicating research needs to research community	
	• Forge better connections between scientists and industry	•••
	• Create awareness (pros and cons)	
	• Promote public relations to "advertise" anotechnologysuccesses to public	•
	• Improve communication of material needs through the value chain	
	• Create on-line magazine like *MIT Technology Review* on nanotechnology subjects	
EH&S		
	• Harmonize and validate toxicology testing protocols	•••
	• Validate *in-vitro* test protocols to correlate to in vivo effect (nanotoxicology)	
	• Employ cross-disciplinary approach to assess issues (e.g., biochemistry, engineering, materials, toxicology, etc.)	
	• EH&Stest protocols validated across multiple labs	
	• Bridge gap between National Institutes of Health (NIH) funding (toxicology) and NSF funding (creation of nanocomposites)	

Exhibit 3.2.3-a. (Continued)

Topic	Solution	Priority

	• Develop tool kits
	  • Collect, review, and disseminate best and most practical EH&Spractices – tool kit
	  • Develop a method for containing nanoparticles – EH&S
	  • Categorize EH&S risk with focus on worker, consumer, general population
	  • Build EH&Sconcepts into product launch model/ business plan
	• Metrology
	  • Identify chemistry/physics properties most useful for risk management decisions along material life cycle (develop methods)
	  • Develop good metrics for determining exposure (environmental, health, etc.)
	  • Develop well-characterized toxicology samples (mono disperse particles)
	• Regulatory
	  • Be proactively involved with regulators and legislators in setting standards for nanomaterials
	• Inform/decide/communicate/influence EH&Sregulations ASAP
		●●●
		
		●●
		
		●

Research		

	• Surface modification – bonding
	  • Binding covalently to the matrix
	• *In situ* formation precipitation, self assembly, gas/water phase dynamics
	• Develop material templates to help control the dispersion
	  • Develop material templates to control structure
	  • Address dispersion issues on an industrial scale
	• Focus on set of composites
	  • Understand how various composite formation methods (e.g., extraction/casting/high shear) impact matrix properties
	• Enhance interaction between synthetic and bi-focused programs on nanotechnologyinterfaces
	• Analyze relationship between nanoscale and discontinuous, amorphous phase
	• Develop high-throughput (lab-on-chip) method to characterize >2 properties simultaneously
		●●●●●●●●
		
		●●●●
		●

Pre-production		

	• Scale up
	  • Low cost raw materials
	  • Consider scale-up issues from product development outset
	  • Incorporate linking/bonding with current materials or processes
	  • Create programs for scale-up facilities
	  • Strategic centers of excellence in nanomanufacturing (nanomaterials and nanotechnology products)
		●●●●●●●●
		●●●●

Exhibit 3.2.3-a. (Continued)

Topic	Solution	Priority

	• Customer needs/value
	  • Collaborate with customers and others in value chain to develop robust systems
	  • Prioritize commercial needs and increase their influence on research funding allocations
	  • Clearly define value of nanotechnology
	• Manufacturing needs/value
	  • Process-focused program, well understood markets and technologies
	  • Provide access to pilot-scale manufacturing equipment for process validation and demos
	  • Develop generic business case plans/templates for risk assessment and sensitivity analysis
	  • Enable low-cost production
	  • Understand process-structure (engineering properties)
	• Modify certification process
	• Develop tool kit with measurement systems for nano- to micro-scale
	  • Develop "labeling" technology to assist measurement of mixing/dispersion
	  • Develop improved metrology/imaging for nanocomposites
	• Gain knowledge of real-time and accelerated aging of well-characterized nanomaterial systems (prioritization needed)
	• Develop robust, real-time, in situ composite characterization and monitoring tools
	• Develop analytical method for non-ideal/complex material systems
		●●●●●●●●
		
		●●●●●
		
		●●●●●

Computing		

	• Petascale computing
	  • Develop modeling tool kits on "nanoHUB.org" to simulate multiple different properties of nanocomposites
	• Model properties and map them across size scales
	  • Engineering and science map
	  • Structural
	  • Strategic targets
		●●●●

3.2.3. Priority Solutions

Cross-industry solutions to address the barriers were discussedand a list of topics identified, as follows:

- Informatics
- Communication
- EH&S
- Research
- Pre-Production (Process, Pilot, Value)
- Measurement
- Computing

EXHIBIT 3.2.3-B. PRIORITY CROSS-INDUSTRY SOLUTIONS NANOTECHNOLOGY-ENABLED COMPOSITES AND MATRICES

Nanomaterials informatics/knowledge base to accelerate progress

Develop methodology to control reactions/bonding of nanoscale particles within a matrix and across interfaces

Build value proposition for nanomanufacturing to meet customer needs

Develop tool kit enabling measurement at nano- tomicro-scale (including dispersion of nanoparticles within a matrix and at interfaces)

Predictably scale up lab-scale processes to manufacturing-scale for production of nanomaterials and products

Optimize manufacturing processes

Coordination of research on nanocomposites and matrices across industry, universities, Government, and Federallaboratories

Science needs within these topical areas were interactively discussed and the importance of possible solutions ranked (see Exhibit 3.2.3-a). While some of the solutions focus on a specific barrier, others apply across multiple barriers and industries.

To hold commercial value for industry, a nanocomposite must offer specialized properties at lower cost than traditional materials or provide multiple new and unique properties for a specific application. These properties must not degrade significantly through the life of the material. Developing these capabilities will require significant research into the interactions of the nanomaterials in the polymer matrix and how these interactions change over time and under various conditions.

From the solutions identified in Exhibit 3.2.3-a, a set of priority solutions was selected and grouped as shown in Exhibit 3.2.3-b. Mini-roadmaps for each solution are shown graphically on the pages thatfollow.

3.3. Separations and Fractionation

Techniques and industry processes related to separations and fractionation of nanomaterials are critical for characterization and stabilization of many materials during manufacturing. Nanomaterial applications in medicine, food production and packaging, chemical pigment production, bio-fuel processing, and the forest industry have specific needs to improve methods and techniques of separations and fractionalizations. Purifications in particular are a targeted emphasis.

Nanotechnology-Enabled Composites and Matrices

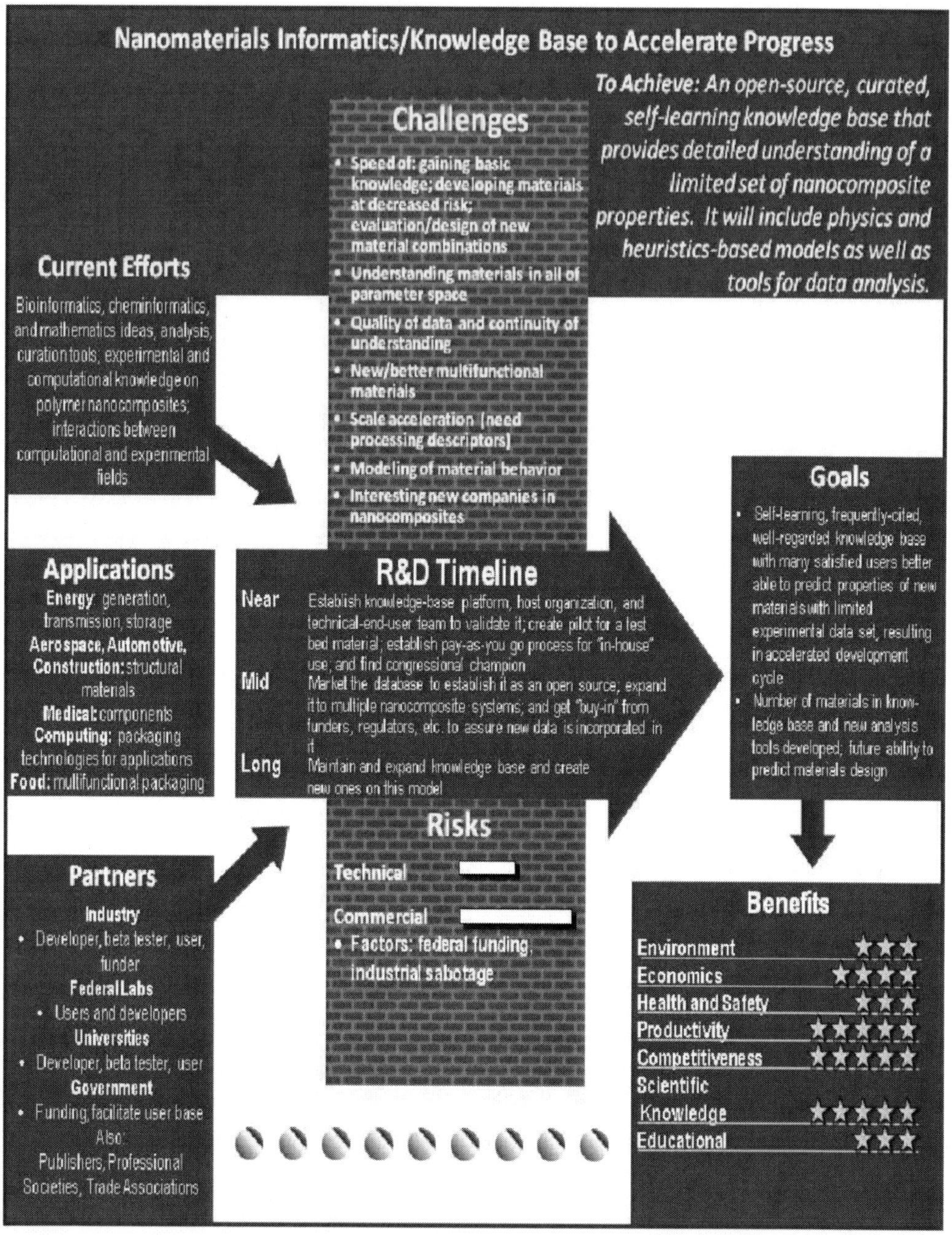

Timeline: Near-term (2009 to 2011) Mid Term (2012 to 2015) Long Term (2016 and beyond)

Priority Solution: Nanomaterials Informatics/Knowledge Base to Accelerate Progress

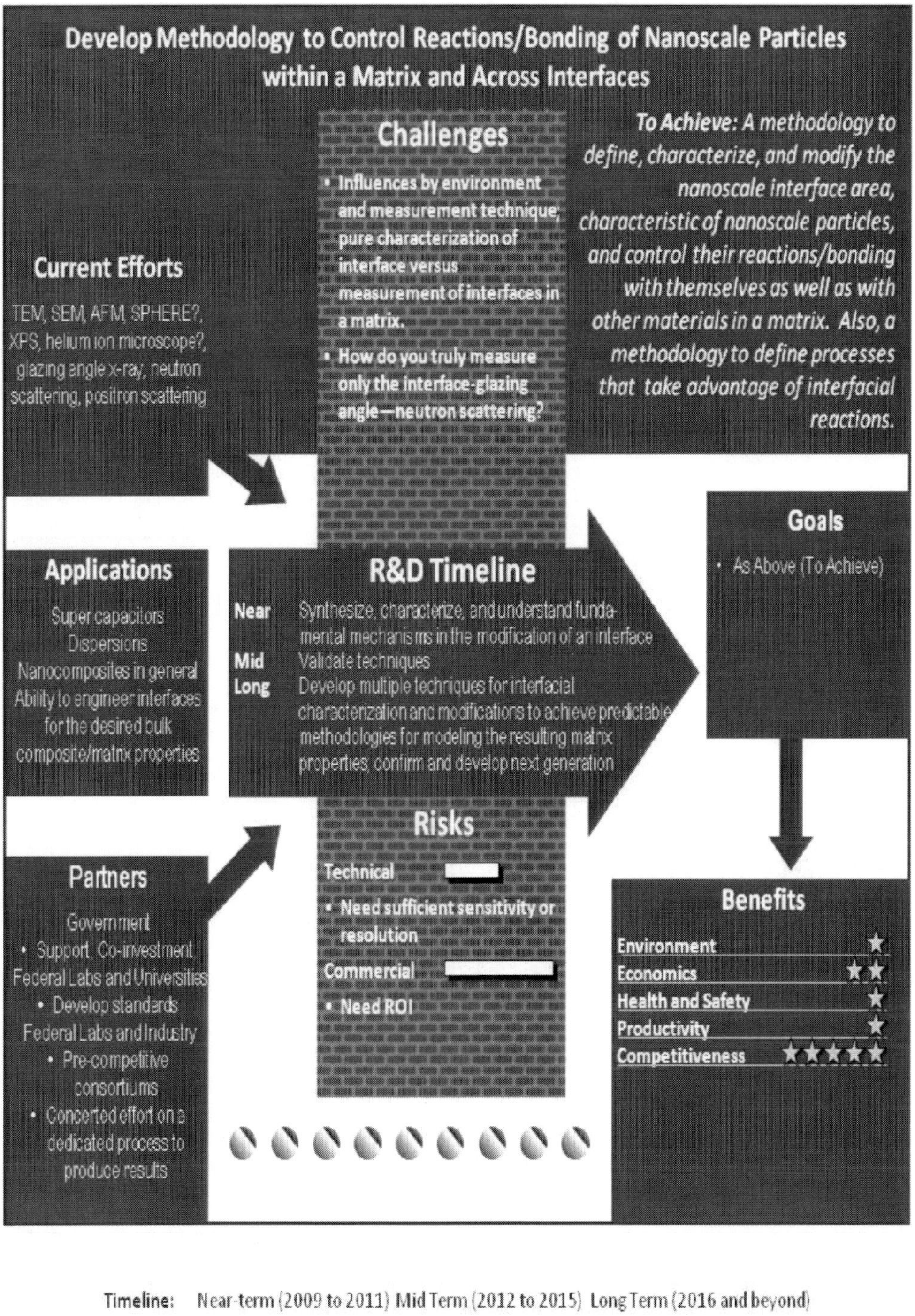

Priority Solution: Develop Methodology to Control Reactions/Bonding of Nanoscale Particles within a Matrix and Across Interfaces

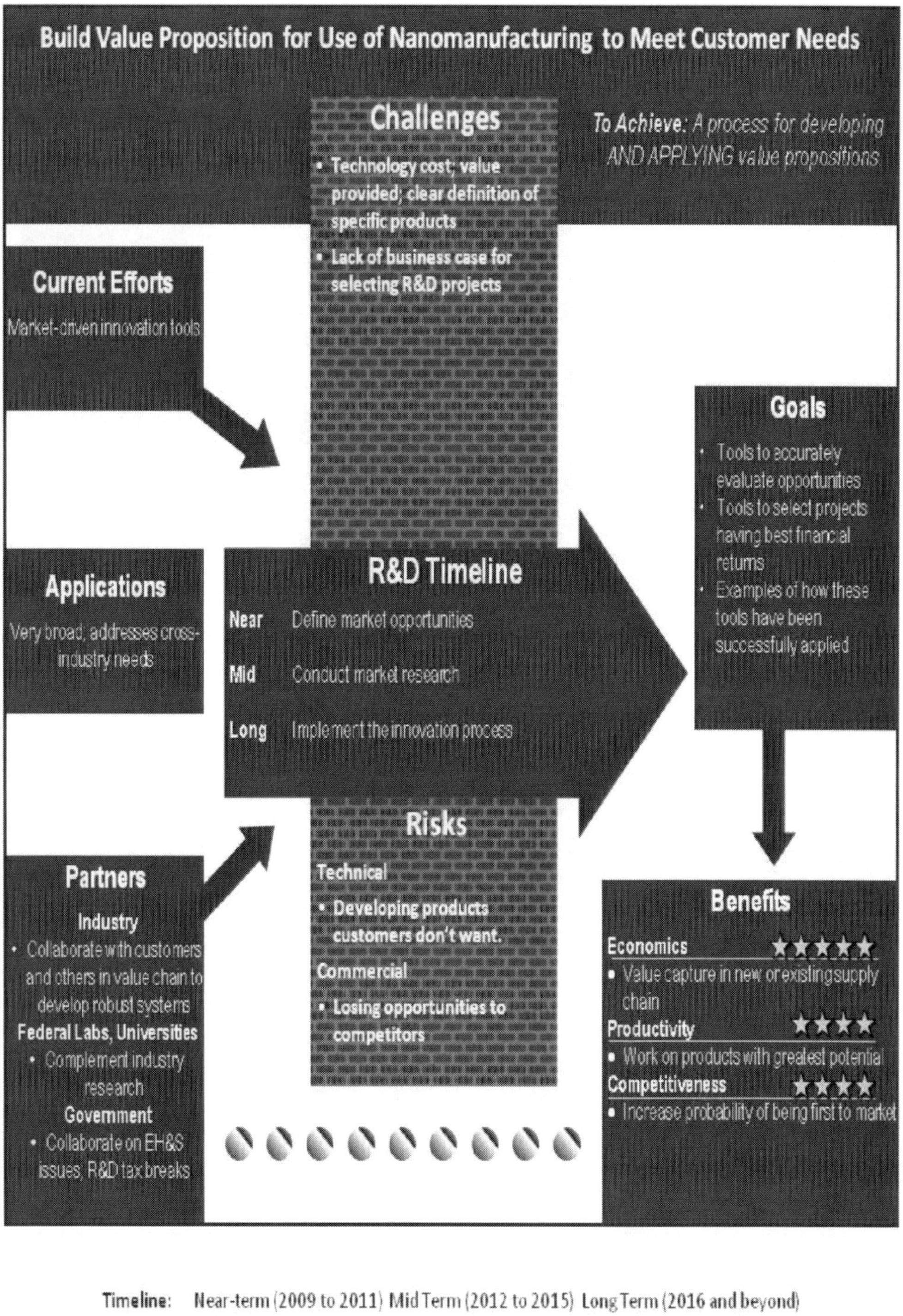

Priority Solution: Build Value Proposition for Use of Nanomanufacturing to Meet Customer Needs

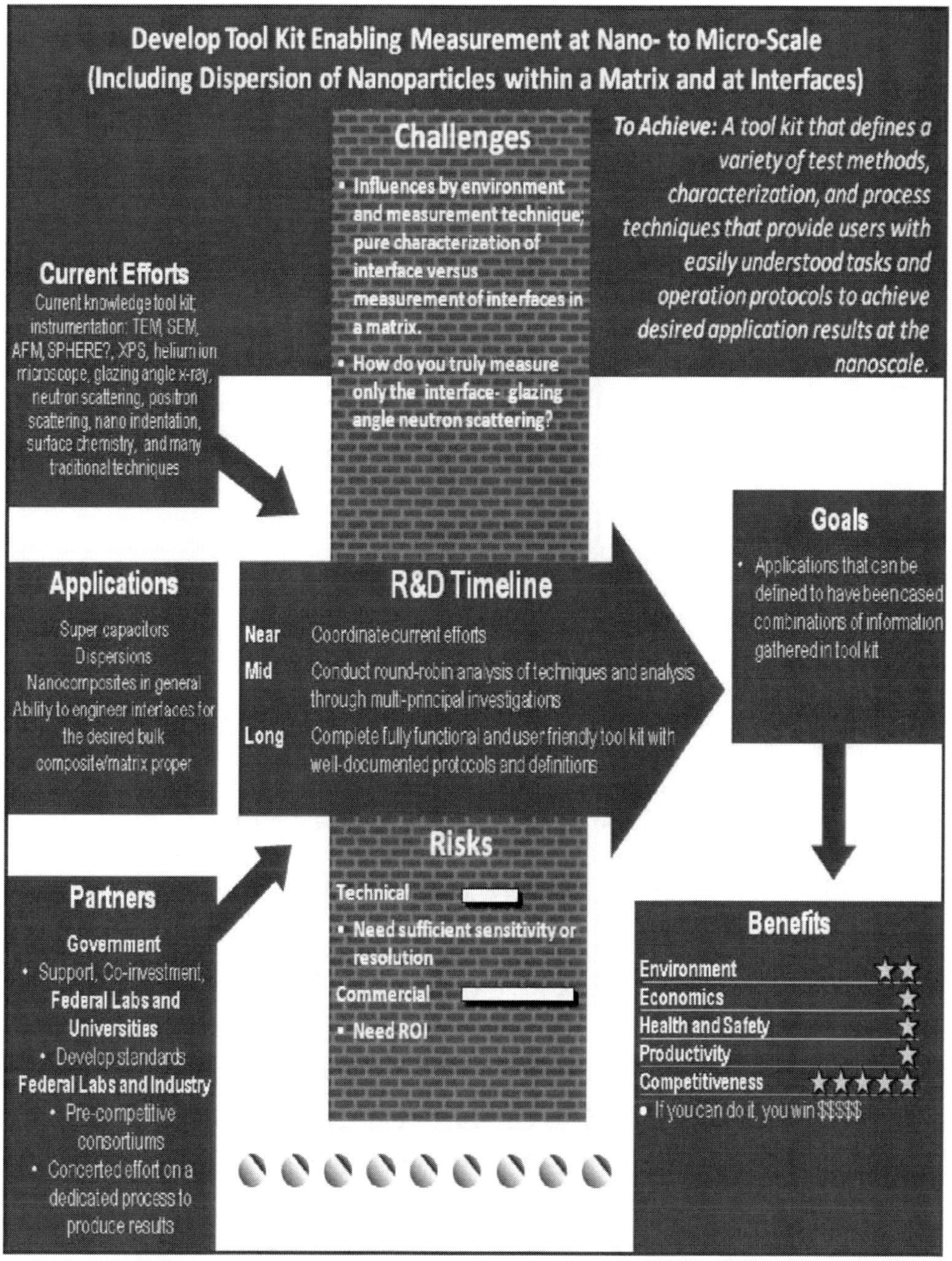

Timeline: Near-term (2009 to 2011) Mid Term (2012 to 2015) Long Term (2016 and beyond)

Priority Solution: Develop Tool Kit Enabling Measurement at Nano- to Micro-Scale (Including Dispersion of Nanoparticles within a Matrix and at Interfaces)

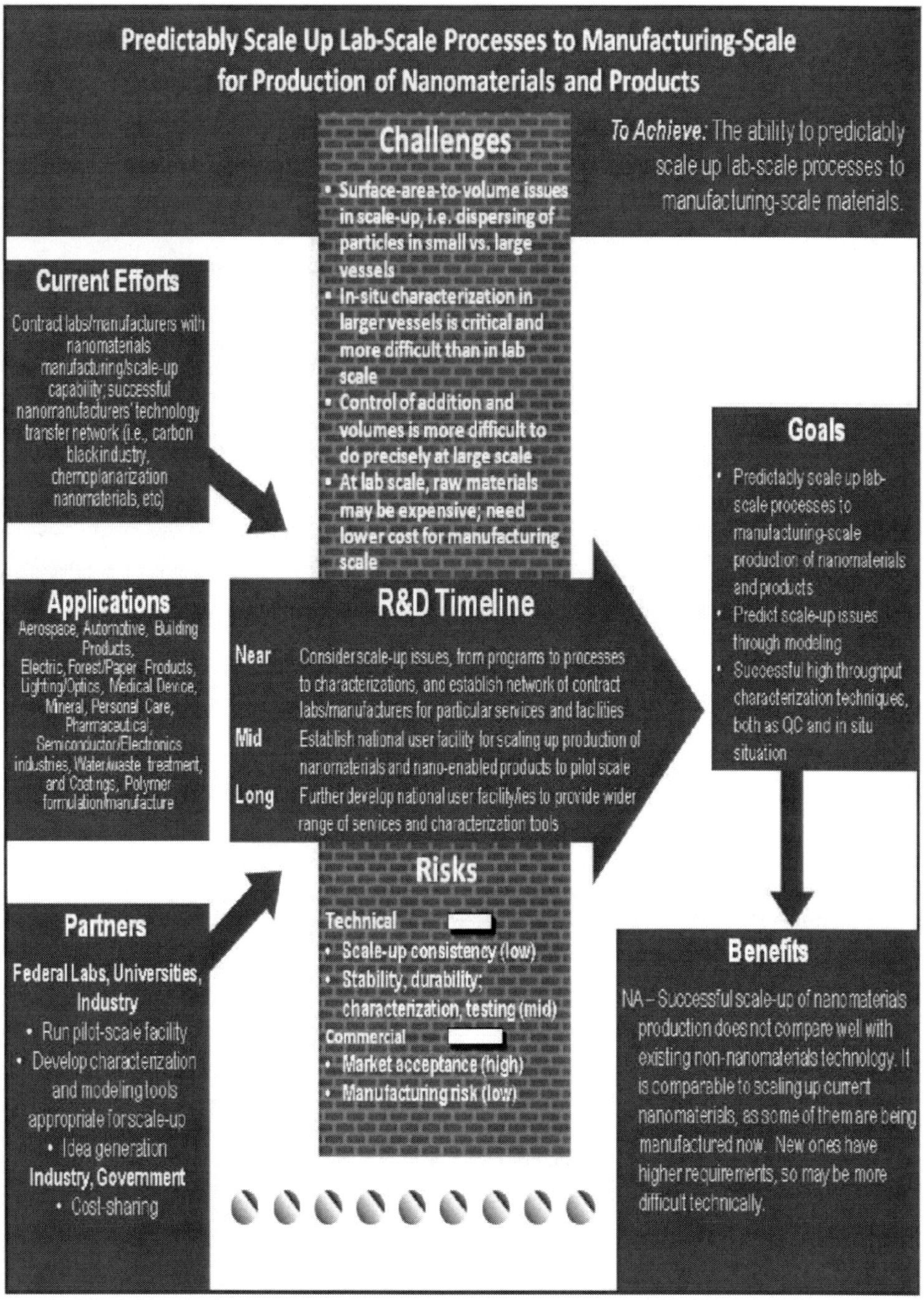

Timeline: Near-term (2009 to 2011) Mid Term (2012 to 2015) Long Term (2016 and beyond)

Priority Solution: Predictably Scale Up Lab-Scale Processes to Manufacturing-Scale for Production of Nanomaterials and Products

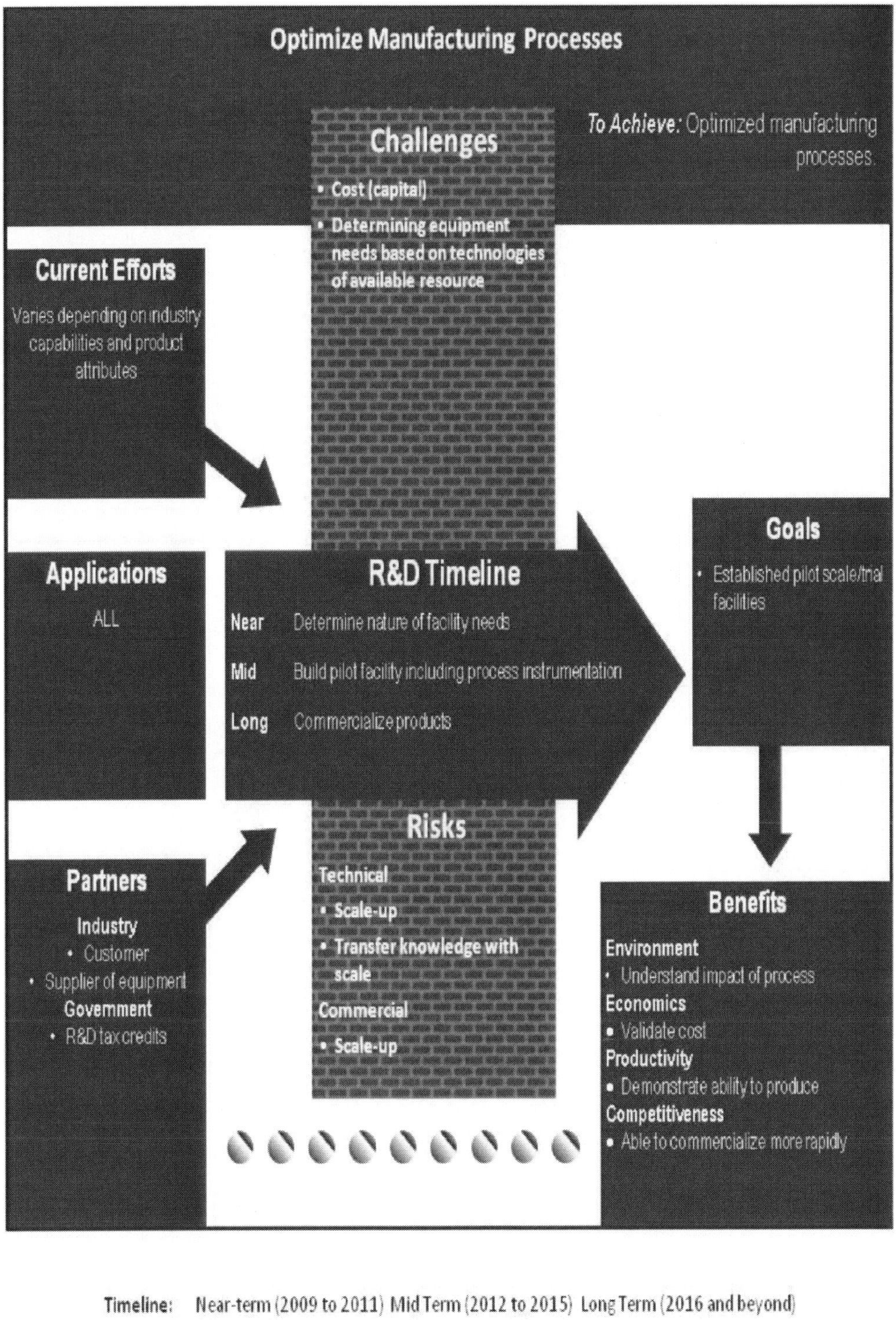

Priority Solution: Optimize Manufacturing Processes

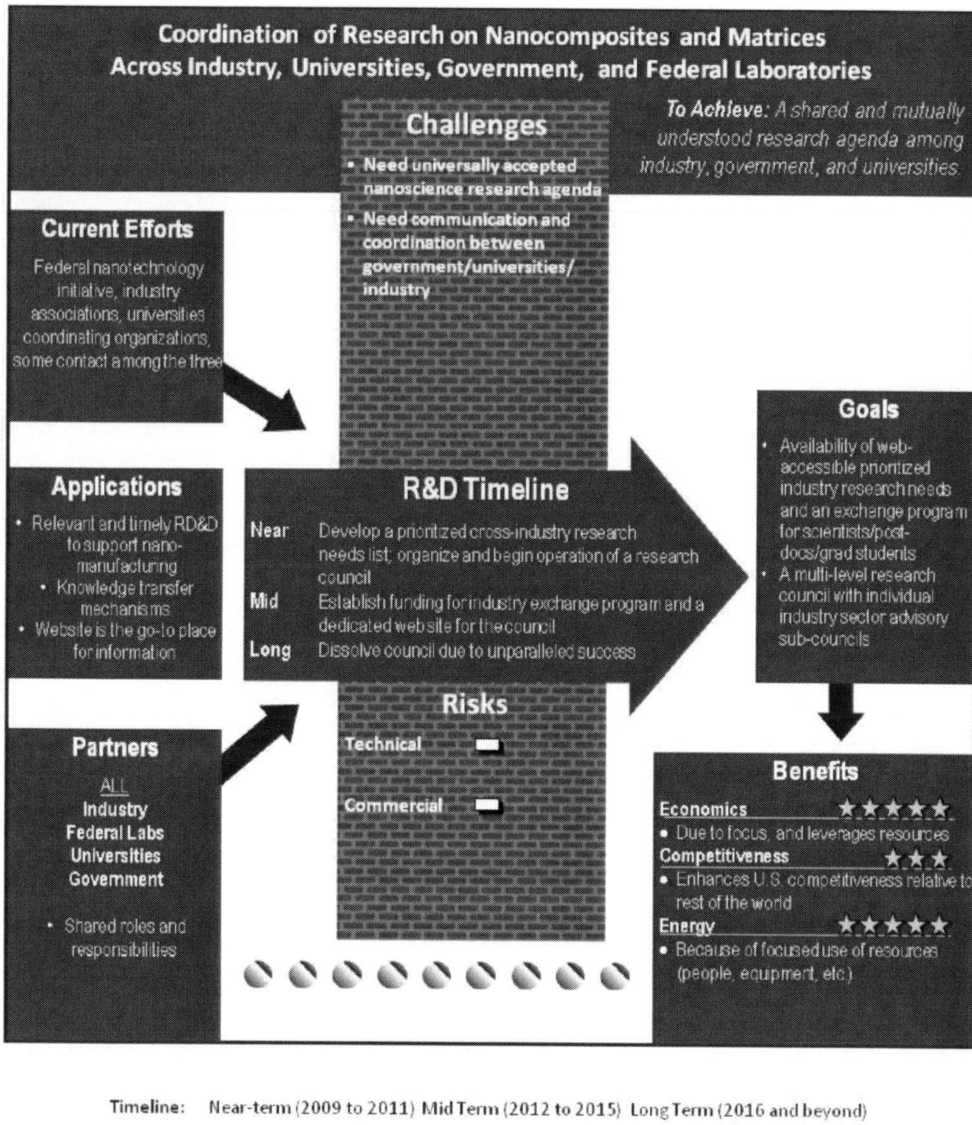

Priority Solution: Coordination of Research on Nanocomposites and Matrices Across Industry, Universities, Government, and Federal Laboratories

This group was tasked with identifying the challenges and prioritizing the needs for advancing separations and fractionation techniques for commercial production of uniform, high quality, stable, and consistent nanomaterials in high volume. It is recognized that separations and fractionation methods are specifically applicable to the production of nanoparticles, which might be functionalized and used for applications such as nanomedicines, or mixed with other materials to form nanocomposites.

Two basic production strategies for nanoparticles were discussed: (1) large volume, batch fabrication, where the application of filtration techniques is used to purify the formulation from undesired byproducts; and (2) emerging microfluidic methods that could produce high-purity formulations that require little purification.

3.3.1. Current Advances

There are a range of techniques that can be applied to the separation and fractionation of nanoparticle formulations. These include filtration, chromatography, electrophoresis, ultracentrifugation, and field flow fractionation. The techniques each have their advantages and disadvantages that are dependent on the type of particle of interest, and the cost, quality, and efficiency of the separation.

Filtration methods have the potential to provide high-speed, high-volume fractionation capability. The basic approach is to set up a series of filtration steps with increasingly smaller pores in order to extract or separate larger-sized materials or debris that would clog the membrane filter for the particles of interest. In addition, each filtration step would yield particle size fractions in the size range of the filters that are upstream and downstream of that step. Examples of state-of-the-art membrane filtration techniques include the use of ordered porous materials (for example, see M. E. Davis, Nature 417, 813, 2002).

The separation of materials by chromatography was first reported by Mikhail Tsvet in 1906. In a chromatographic separation the material of interest is dissolved in a solvent (mobile phase) gas or liquid, and is passed over a solid adsorbent (stationary phase), which is typically packed in a chromatographic column. The differing affinities between the material, the solvent, and the solid adsorbent will separate the material of interest into its constituent parts. One advantage of this technique is that it is well established and has been expanded for application to a host of materials. A disadvantage is that it is not applicable for high-volume production. Techniques include normal phase (NP-HPLC), reverse phase (RP-HPLC), size-exclusion or gel permeation (SEC or GPC), and ion exchange (IEC), and fast protein liquid chromatography (FPLC).

Another well-established separation method is electrophoresis, discovered by Ferdinand Friedrich Reuss in 1806. Electrophoresis is a primary technique for biomolecular separations because it is powerful yet reasonably easy and inexpensive to implement. It can be carried out in a rectangular slab of material, a capillary tube, and in planar microchip formats. Electrophoretic separation techniques include free-solution and gel electrophoresis, micellar electrokinetic, and isoelectric focusing. This technique can yield high-quality separations of materials for laboratory analysis but is also not amenable to high-volume production.

Session Scope

- Identify what separation, purification, and fractionation issues are critical to nanotechnology
- Define what kinds of separations and fractionalizations are used to get formulations needed for products and current limits
- Identify challenges and barriers for improving current separations and fractionations to improve yields

Separations have also been demonstrated using a technique called analytical ultracentrifugation, invented in 1923 by Theodor Svedberg. The separation is carried out in a device called an ultracentrifuge, which is optimized for spinning a rotor at very high speeds that are capable of generating acceleration as high as 1,000,000 G. The sample being spun can be monitored in real time using ultraviolet light absorption and/or an interference optical refractive index sensitive system. A hazard associated with analytical ultracentrifugation is the tremendous rotational kinetic energy where a catastrophic failure of the rotor is a serious concern.

The last method of common use is separation by field flow fractionation (FFF), discovered in 1965 by J. Calvin Giddings. An excellent review of this technique can be found in the reference: J.C. Giddings, Science, 260 (1993), 1456. The separation method is based on the differing retention of materials in a stream of liquid flowing through a thin channel while applying a force (by centripetal, electric, or magnetic field) at right angles to the flow. The technique will allow a sample to elute through the system and separate by size. This technique has been demonstrated to give high-performance separations of nanoparticles but is not scalable to high-volume production at this time.

3.3.2. Key Barriers

Specific barriers for improving current separations and fractionations to improve yields identified in the workshop are detailed in Exhibit 3.3.2-a. In summary, these include:

Manufacturability. The main challenges and problems in separations and fractionation are in the area of manufacturability.[23] A consensus was reached that improved in-line measurements and characterization of nanoparticles was needed. Filtration techniques typically require a multiple step process. In-line measurements at each filtration step would yield improved understanding of the process so that it could be used to develop models to optimize the process for high throughput and lowered cost. In-line measurements were also desired for the microfluidic techniques, which could lead to improved understanding of how nanoparticles form and could drive massively parallel production techniques to bring throughput to commerically viable volumes.

Characterization. Key characterization issues that present challenges are simultaneous sorting of size, shape, and charge; and temperature and chemical compatibility. It was recognized that the ability to not only characterize nanoparticle size, but to also simultaneously determine shape and charge, would be a highly useful improvement in characterization.

Purification and Removal of Non-Nanomaterials. Purification is a key step in the development of nanomaterials with the desired properties and functionality. Purification and fractionation first take place in the removal of non-nanomaterials. The main issues that need to be addressed in this area are preparative purification, sample preparation, and the removal of debris. There are also challenges in assuring the purity of the material produced for use in final products, including the sensitivity of the process and selectivity.

Exhibit 3.3.2-a. Key Barriers for Separations and Fractionation

Barriers	Needs	Priority
Measurement Methods	Calibration and reference materials	●●
	Sample collection method; getting enough of the item of interest	
	Measuring the metric of interest to neededtolerance	
	Instrument selectivity	
	Instrument sensitivity	
	Detection limit	
	Precision	
Removal of Non-Nanomaterials	Preparative purification	●●●
	Sample preparation removing debris	●●
	Soluble fraction – separate	
	Semi-preparation, application/purification	
Manufacturability	In-line measurements and visualization	●●●●●●
	High-throughput methods	●●●●●●
	Efficiency	●●●
	Avoid aggregation, recognize change in extract	●●●
	In-line characterization	●
	Filtration rates	●
	Making stable dispersions	●
	Reproducibility, consistency	●
	Quantity (large) processing (scale-up)	●
	Mature manufacturing, large scale, process availability	
	Longevity, i.e., get enough product made without blocking	
	Yield	
	Affordability of existing filtration & separation methodologies	
Purification	Probes selectivity	●
	Sample and final product purity	
	Process sensitivity	
	Purity of raw material produced	
Characterization	Simultaneous sorting of size, shape, and charge	●●●●●●
	Temperature and chemical compatibility	●●
	Size determination in water	
	Defined attributes for separation/fractionation	
	Verification of separation/fractionation	
	Matrices (air, liquid, biofluid)	
	Compatibility – no fouling/blocking	
	Chemical composition	
	Modeling the system	

Measurement Methods. Key measurement issues encompass the ability to measure the properties of interest with enough precision, selectivity, sensitivity, and tolerance in the manufacturing environment. Proven calibration techniques that are reliable for production and reference materials are also needed to ensure reproducibility.

3.3.3. Priority Solutions

Cross-industry solutions to address the barriers were discussedand a list of topics identified, as follows:

- Avoid aggregation; recognize change in extract
- Efficiency
- Preparative purification
- Modeling
- High throughput methods
- In-line measurements and visualization
- Simultaneous sorting of size, shape, charge

Research needs within these topical areas were further explored and the importance of possible solutions ranked (see Exhibit 3.3.3-a). These solutions are mostly unique to separations and fractionation, but many could be appliedacross multiple industries.

Exhibit 3.3.3-a. Cross-Industry Solutions for Separations and Fractionation

Topic	Solution	Priority
Avoid Aggregation; Recognize Change In Extract		
•	Plate reader high throughput approach	
•	Agitation/shaking, additives/surfactants, flow aids	
•	Use micro-bubble reactor to avoid aggregation	
Efficiency		
•	Sharp cutoffs so that multiple passes are not needed	●
•	Easy to clean	●
•	Filter sized correctly for particle size required	
Preparative Purification		
•	Filter or centrifuge sized correctly to remove debris, junk, contamination	
•	Particle size required (nm)	
Modeling		
•	Models of modules for system design. Optimization by modeling – solvent, charge, interaction	●●●●●●●●
•	Sizing determined by modeling – feed, reaction and reaction profile, required final product	
•	Model for efficiency – feed, reacti on profile, required final product	
•	Model correctly – feed, reaction, final product	
High Throughput Methods		
•	High-throughput production/scalable	●●●
•	Investigate massively parallel microfluidic methods	●●●
•	Combine synthesis and separation to increase process throughput/efficiency	●●

Exhibit 3.3.3-a. (Continued)

Topic	Solution	Priority
•	Centrifugation filtration chromatography	•
•	Investigate use of rotatingpacked beds for materials separation	
In Line Measurements and Visualization		
•	Adapt analytical methods to in-line. Filter to nm range, charge capture release <50 nm, side stream/sample port, zeta potential laser	••••••••
•	Use ultrasound attenuation spectra in line for measurement/control	••
•	Dynamic light scattering (DLS)	••
•	3D measurement and visualization	•
•	Develop microfluidic method and measurement techniques	••
•	Automated microscopic sampling and analysis	
•	Side stream filter to <50nm, laser, zeta, microfluidic channel, print out real time on size distribution	
Simultaneous sorting of Size, Shape, Charge		
•	Standard reference materials, size, shape, charge, chemistry	••
•	Modify electro-zone sensing for size, shape, and charge sorting	
•	Investigate nanofluidic methods	
•	In-line mass spectroscopy, enhanced mass spectroscopy	
•	Microfluidics with detection capabilities	

From the solutions identified in Exhibit 3.3.3-a, a set of priority solutions were selected and grouped as shown in Exhibit 3.3.3-b. Mini-roadmaps for each solution are shown graphically on the pages that follow.

Exhibit 3.3.3-b. Priority Cross-Industry Solutions Separations and Fractionation

Develop parallel microfluidicmethods

Build high-throughput production/scalable methods

Enable modeling to predict process/material characteristics

Develop in-line measurements and visualization: transfer laboratory analytical methods to in-line

SEPARATIONS AND FRACTIONATION

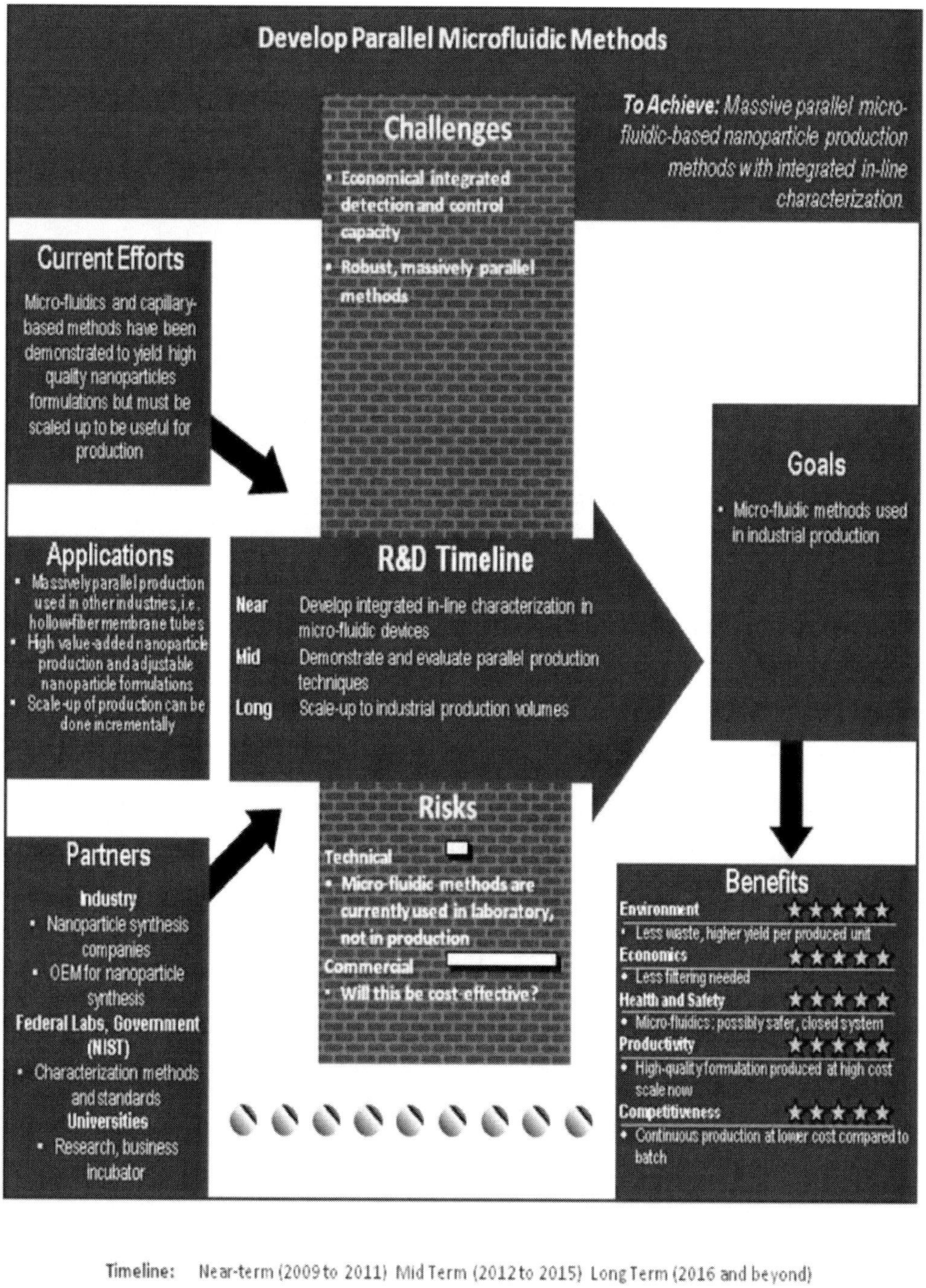

Priority Solution: Develop Parallel Microfluidic Methods

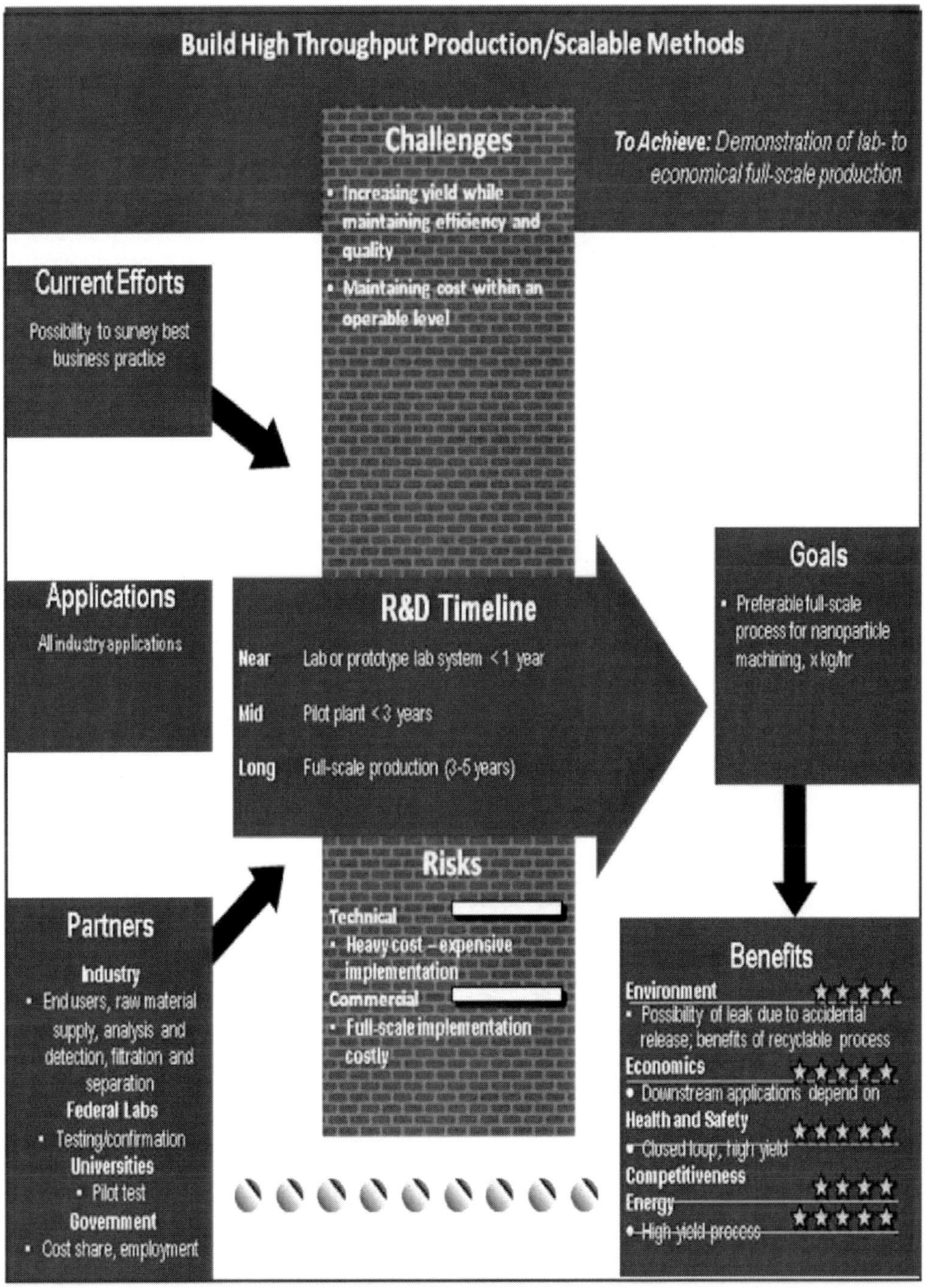

Timeline: Near-term (2009 to 2011) Mid Term (2012 to 2015) Long Term (2016 and beyond)

Priority Solution: Build High Throughput Production/Scalable Methods

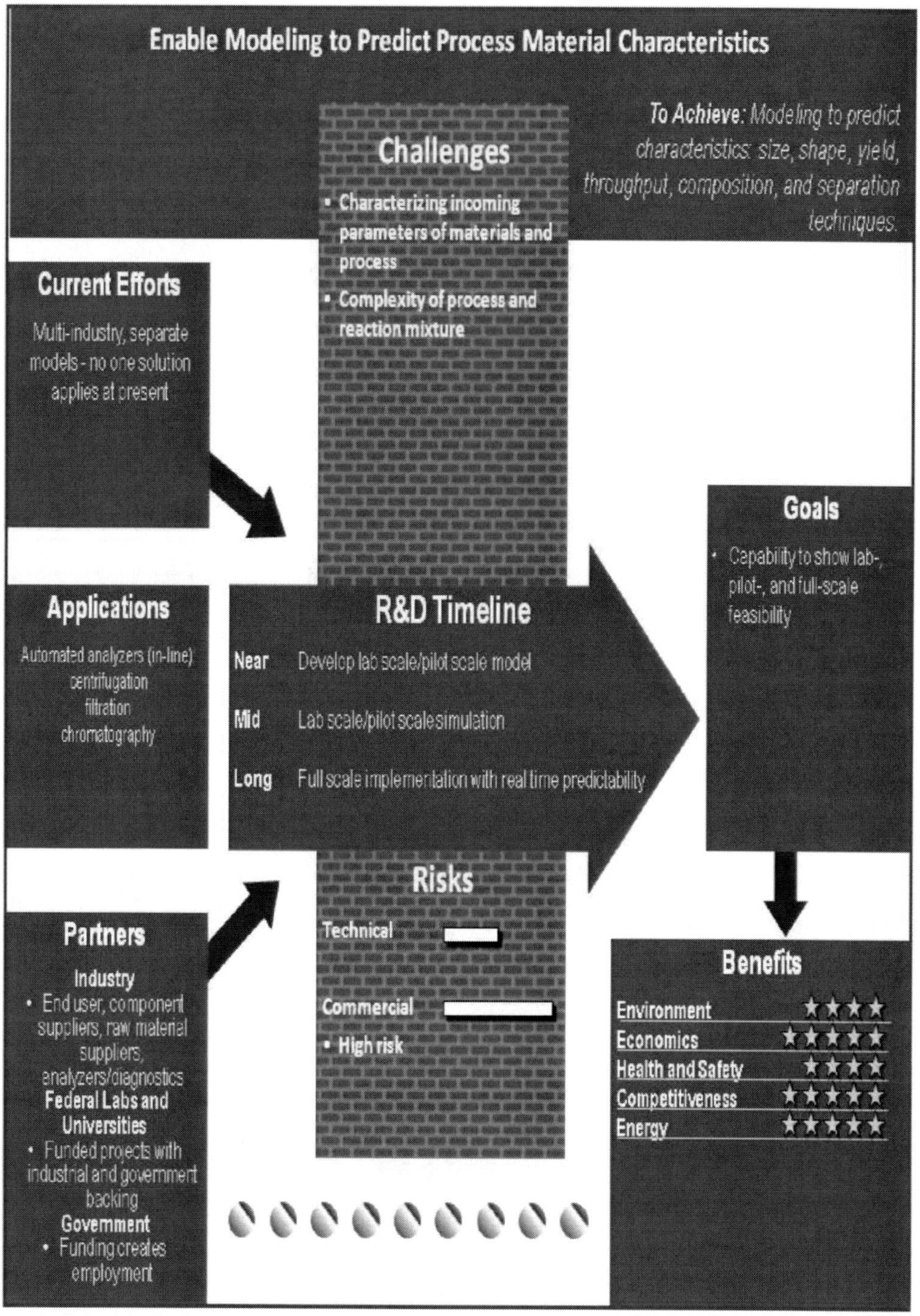

Timeline: Near-term (2009 to 2011) Mid Term (2012 to 2015) Long Term (2016 and beyond)

Priority Solution: Enable Modeling to Predict Process/Material Characteristics

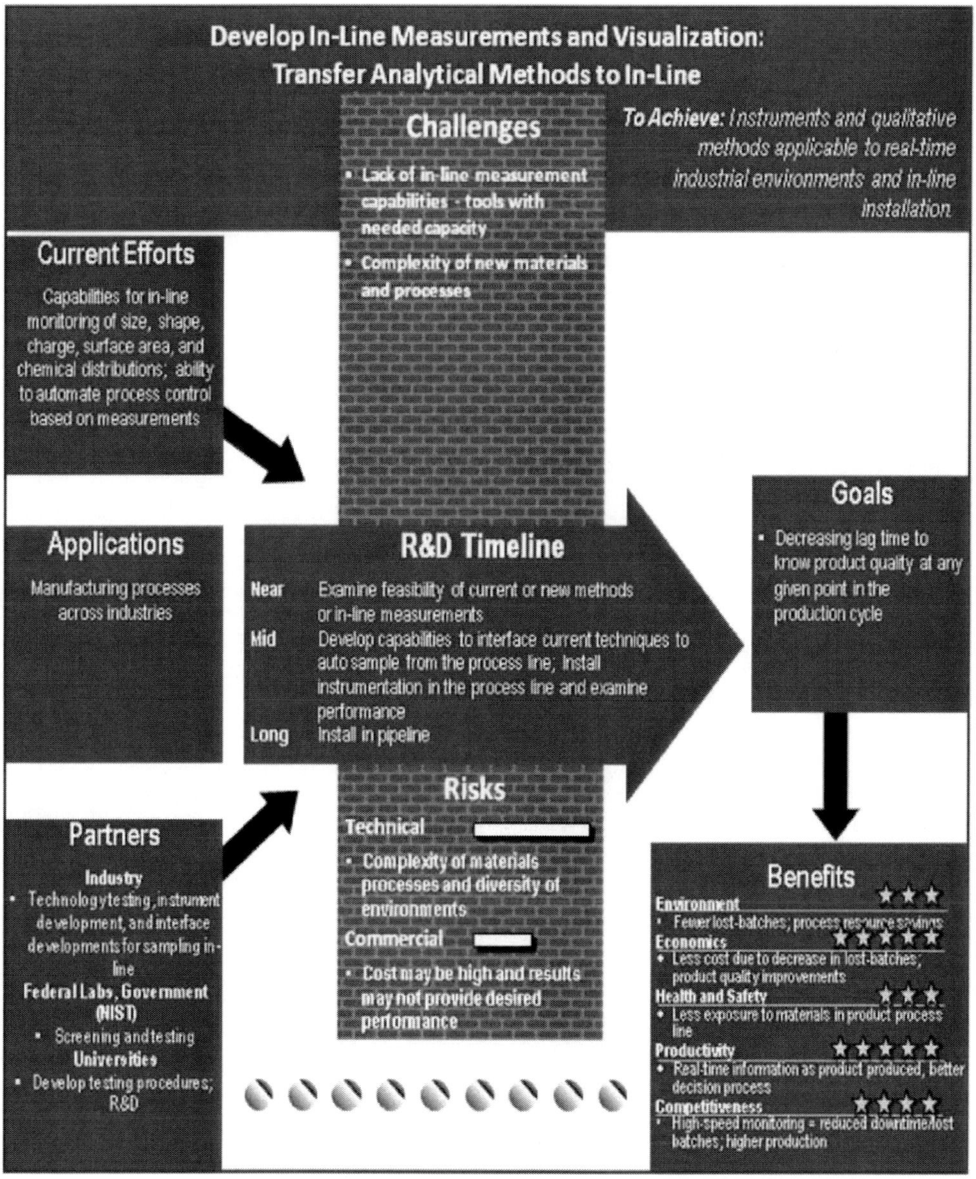

Timeline: Near-term (2009 to 2011) Mid Term (2012 to 2015) Long Term (2016 and beyond)

Priority Solution: Develop In-line Measurements and Visualization: Transfer Laboratory Analytical Methods to In-Line

4. The Path Forward

Identifying and better understanding the cross-industry challenges is a step toward coordinating research, disseminating common knowledge and that could serve all industries, and improving the knowledge foundation and data infrastructure needed for cost-effective, world-competitive nanomanufacturing. Currently, multiple industries and sectors are producing nanotechnology roadmaps, often with similar or cross-cutting needs. Multiple databases of information are beginning to emerge, along with technical successes in commercial nanotechnology. Along with these developments is a strong consensus on the need for evaluating the environmental, health, and safety impacts of nanotechnology-based products. Modeling, measurement, and process control, based on high throughput measurement, have been established as critical parameters to the safe commercial utilization of nanotechnology on a broad scale.

While there is a large Federal investment in nanotechnology research, the challenge remains in prioritizing what needs to be done, and how to leverage the outcomes (gaining the most benefit for dollars spent). Many government agencies are making a push toward cross-industry programs to expand the reach of R&D. The need for coordination in nanomanufacturing R&D and information dissemination across the public and private sectors has never been stronger, and is critical to attaining and maintaining global leadership.

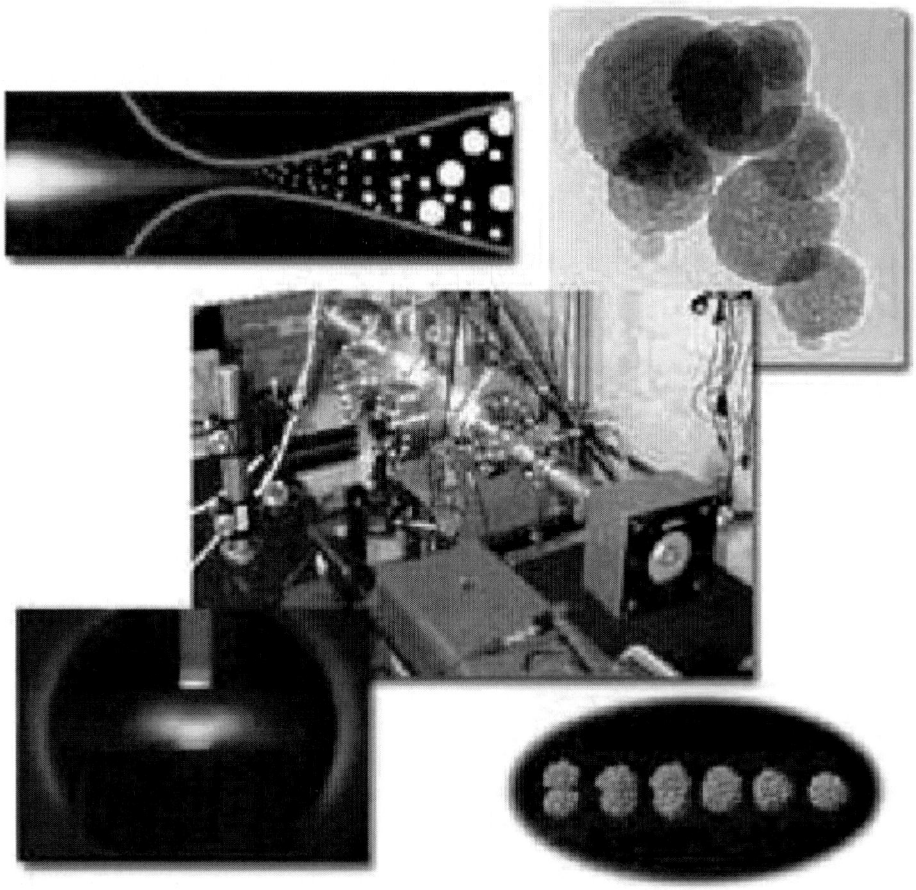

Figures courtesy of The University of Maryland (http://www.enme.umd.edu/cnmm/)

As the results of this workshop illustrate, there are numerous cross-industry challenges related to nanomanufacturing that could begin to be addressed through a concerted, integrated, public-private effort. The following guiding concepts were considered to be an essential foundation for the framework for a coordinated effort:

- Enable cross-fertilization and identification of best practices using currently available science and technology to deliver short-term impact,
- Define collaborative research programs that cross industrial sectors, government agencies, and academic disciplines to address the more difficult challenges and long-term needs,
- Identify challenges that are appropriate for Federal and/or industrial funding, and lay the groundwork for formation of consortia and multi-organizational R&D projects to achieve the desired outcomes,
- Establish a reference data infrastructure that links nanoscale structure and composition to properties and performance to enable more effective R&D, advance fundamental understanding, and the development and validation of reliable predictive models, and
- Establish a knowledge database of work conducted at universities and national labs that captures structure/manufacturing process/performance relationships, as well as predictive capabilities of models and simulations.

Ideas for creating a coordinating nanomanufacturing group were presented and explored. One important idea that emerged was an umbrella alliance that would be industry-driven (industrial decision makers with long-term commitment, trade associations, individual companies) and government-facilitated and government-supported. The purpose of this alliance would be to help consolidate multi-industry research needs, map existing research and capabilities, consortia, and initiatives, and build a nanomanufacturing knowledge base in targeted areas.

Implementing an umbrella alliance will require the development of planning committees and an organizational charter. Tentatively this would include an executive committee, advisory committee, administrative group, and technical committees/subcommittees – all with members from both the public and private sectors. Discussions have taken place post-workshop to build an Umbrella Alliance for Nanomanufacturing. As this organization evolves, it is expected that the benefits of a coordinated approach to this critical area of nanotechnology will begin to emerge.

APPENDIX A. LIST OF PARTICIPANTS AND CONTRIBUTORS

Clare Allocca
NIST

Mohamad Al-Sheikhly
University of Maryland

Gary Fletcher* **
Becton Dickinson

Martin Fritts
Nanotech Characterization Lab

George Andrews**
General Motors

Adra Baca
Corning, Inc.

Kathryn Beers
NIST

Ron Brown**
Agenda 2020 Technology Alliance of AF&PA

Millie Calistri-Yeh
Becton, Dickinson and Company-BD Medical

Anne Chaka, Chair
NIST

Shaochen Chen**
NSF

Meng-Dawn Cheng
Oak Ridge National Laboratory

Gerard Closset, Co-Chair
Agenda 2020 Technology Alliance of AF&PA

Dan Coughlin**
Sappi Fine Paper

John Cowie, Co-Chair
Agenda 2020 Technology Alliance of AF&PA

Elizabeth (Betsy) Davies**
Agenda 2020 Technology Alliance of AF&PA

Travis Earles
Office of Science and Technology Policy

John Festa**
Agenda 2020 Technology Alliance of AF&PA

Michael Gaitan* **
NIST

Michael Garner**
Intel

Charlie Gause
Luna Innovations

Robert Gelman**
Hercules Inc.

Charles Geraci
NIOSH

Jeff Gilman
NIST

Angela Hight Walker
NIST

Gale Holmes*
NIST

Ehr Ping Huangfu**
Department of Energy

Keith (Dan) Humfeld
The Boeing Company

Phil Jones**
Imerys

Philip Lippel
National Nanotechnology Coordination Office

Shawn Lofstrom
Sloan Biotechnology Industry Center
University of Maryland

* Session Chair
** Organizer

Terry Lynch
NIST
Bernadene Magunson
Cantox Health Sciences International

Christine Mahoney*
NIST

Mohan Manoharan
GE Global Research

Steven Masia* **
SAPPI Fine Paper

Jeffrey Meth

DuPont

Robert Moon
U.S. Forest Service/Purdue University

Richard Morris
Pall Corporation

Benjamin Naden
Imerys Minerals Ltd.

World Nieh* **
U.S. Forest Service

Evgenia Pekarskaya
Lux Research

Michael Postek, Co-Chair
NIST

Dianne Poster, Co-Chair
NIST

Karthik Ramaratnam
International Paper

Timothy Rials
University of Tennessee

Nora Savage**
EPA

Linda Schadler
Rensselaer Polytechnic Institute

Paul Schulte
National Institute for Occupational Safety and
Health

John Simonsen
Oregon State University

Srini Sridharan
Becton, Dickinson and Company-BD Medical

Aleksandr Stefaniak
NIOSH

Treye Thomas
US Consumer Product Safety Commission

Albert Tsai
FMC Bioplymer

Mark Tuominen
University of Massachusetts

Wyatt Vreeland
NIST

Theodore Wegner* **
U.S. Forest Service

Michael Wolcott
Washington State University

Xingcheng Xiao
GM R&D Center

Dierk Roessner*
Wyatt Technology Europe GmbH

Nigel Sanders
Specialty Minerals Inc.

* Session Chair
** Organizer

APPENDIX B. WORKSHOP AGENDA

Cross-industryIssues in Nanomanufacturing Workshop National Nanotechnology Initiative Affiliated Workshop Workshop Agenda NIST Campus, Gaithersburg, MD, USA 20899 May 20 – 22, 2008

Tuesday, May 20: *"The Problem" Convener: Anne Chaka (NIST)*		
7:00 -7:30 am	**Continental Breakfast**	
7:30 -8:45am **Plenary Session**	***"Set the stage for the workshop: What are we intending to accomplish and why and what are the Deliverables"***	
	7:30 – 7:50am	**Introductions, antitrust, purpose of workshop, deliverables, workshop logistics, facilitated breakout sessions orientation:** Anne Chaka (NIST)
	7:50 –8:00 am	**Working with NIST:** Terry Lynch (NIST Office of Technology Partnerships)
	8:00 –8:45 am	**Overview of outcomes and follow-up from recent relevant Nanotechnology conferences and workshops:** Mike Postek (NIST)
8:45 am -12 noon **Morning Plenary Speakers**	8:45-9:10 am	**Nanomanufacturing Overview:** Mohan "Mano" Manoharan (Manager, Surface and Coatings Technology Laboratory at GE Research)
	9:10-9:35 am	**Forest Products:** Steve Masia (Sappi Fine Paper)
	9:35-10 am	**Automotive:** Xingcheng Xiao (Senior Researcher, General Motors R&D)
	10:00— 10:15 am	Break
	10:15— 10:40 am	**Pharmaceuticals and Medical Devices:** Gary Fletcher, (Advanced Technology Leader / R&D, Becton Dickinson)
	10:40— 11:05 am	**Chemicals and Semi-Conductors:** Mike Garner (Intel)
	11:05— 11:30 am	**Aerospace:** Keith McIver (Boeing)
	11:30 am— 12:00 noon	**Food Science:** Bernadene Magnuson (Senior Scientific & Regulatory Consultant, Cantox Health Sciences International)

(Continued)

12:00 –1:15 pm **Lunch and EH&S Speaker**	**Speaker:** Paul Schulte (Director of the Education and Information Division at NIOSH, and coordinator of the agency's NanotechnologyResearch Program)	
Afternoon Plen-ary Speaker 1:15-1:140	Mohamad Al-Sheikhly, Department of Materials Science and Engineering, University of Maryland, College Park "Challenges in the Synthesis of Nanomaterials for a Wide Variety of Applications"	
1:40 -3:15 pm **Concurrent breakout groups A, B, and C**	Group A	**Surface/Interfaces and Non-BondedInteractions of Nanomaterials** Session Chairs: Gary Fletcher (Becton Dickinson); Christine Mahoney (NIST)
	Group B	**Nanotechnology-enabled Composites and Matrices** Session Chairs: Steve Masia, (Sappi); Ted Wegner and World Nieh (USDA Forest Service); Gale Holmes (NIST)
	Group C	Separations and Fractionation Session Chairs: Michael Gaitan (NIST); Ron Rossner (Wyatt Technology Corp.)
3:15 –3:30 pm	**Break**	
3:30 –4:30 pm	**Breakout groups prepare report**	
4:30 –5:00 pm **Joint session**	**Highlights of breakout reports.** Each presentation is at most 10 minutes (5 min presentation and 5 minutes Q&A)	
Dinner	**Dinner at Holiday Inn.** Speaker:Michael Holman (Lux Associates)	
Wednesday, May 21: *"The Solution" Convener: Michael Postek (NIST)*		
7:00 -8:00 am	**Continental Breakfast**	
8:00 –12 noon **Expert speakers to address the nanomaufacturing areas of focus**		
	8:00-8:05 am	**Call to Order and Announcements**
	8:05-8:40 am	Nanomanufacturing Overview: Mark Tuominen (University of Massachusetts-Amherst)
	8:40-9:10 am	**Separations/Fractionation:** Wyatt Vreeland (Biochemical Science Division, NIST)
	9:10-9:40 am	**Composites:** Jeff Gilman (NIST), Huber
	9:40-10:10 am	**Surfaces/Interfaces:** Linda Schadler (RPI)
	10:10-10:30 am	Break
10:30 am – 12 noon **Concurrent breakout sessions Groups A, B, and C**	Group A	**Surface/Interfaces and Non-Bonded Interactions of Nanomaterials** Session Chairs: Gary Fletcher (Becton Dickinson); Christine Mahoney (NIST)
	Group B	**Nanotechnology-enabled Composites and Matrices** Session Chairs: Steve Masia, (Sappi); Ted Wegner and World Nieh (USDA Forest Service); Gale Holmes (NIST)
	Group C	**Separations and Fractionation** Session Chairs: Michael Gaitan (NIST); Ron Rossner (Wyatt Technology Corp.)
12 noon –1:00 pm	**Lunch. Speaker: "The United States Measurement System and Measurement Needs for Nanotechnology and NanoEH&S"** Clare Allocca(NIST)	
1:00-3:15 pm	**Groups A, B and C reconvene**	
3:15 –3:30 pm	**Break**	

(Continued)

3:3 0 –4:30 pm	**Breakout groups prepare report**	
4:30 –5:00 pm **Joint session**	**Highlights of breakout reports.** Each presentation is at most 10 minutes (5 min presentation and 5 minutes Q&A)	
5:00-5:30 pm	**"The Way Forward: Potential interaction models for collaboration and cross-fertilization"** Anne Chaka (NIST) and Mike Garner (Intel)	
Dinner	**Picnic Barbeque at Smokey Glen Farm.**	
Thursday, May 22: *"The Way Forward" Convener: Dianne Poster (NIST)*		
7:00 -7:00 am	**Continental Breakfast**	
8:00 to 9:00 am **Session Chairs present the Deliverables from their session.**	8:00 – 8:20am Group A	**Surface/Interfaces and Non-Bonded Interactions of Nanomaterials.** Session Chairs: Gary Fletcher (Becton Dickinson); Christine Mahoney (NIST)
	8:20 –8:40 am Group B	**Nanotechnology-enabled Composites and Matrices.** Session Chairs: Steve Masia, (Sappi); Ted Wegner and World Nieh (USDA Forest Service); Gale Holmes (NIST)
	8:40 – 9:00am Group C	**Separationsand Fractionation.** Session Chairs: Michael Gaitan (NIST); Dirk Rossner (Wyatt Technology Corp.)
9:00-9:15 am	Break	
9:15-11:45 am Next steps	***"Set the stage for the next steps: What are we intending to acco-mplish next and howwho does what, when"*** ***Session Chairs: Gerard Closset (AF and PA), Mike Garner (Intel), Anne Chaka (NIST)***	
	• Follow-up action plan	
	• Model for collaboration	
	• Timeline	
11:45 am – 12 noon	**Concluding remarks. Workshop ends. (Anne Chaka)**	
12:00 noon	**Lunch** (Optional)	
12:30-2:30 pm	**NIST tour** (Optional)	

APPENDIX C. LIST OF PRESENTATIONS

The following presentations were made at the workshop. Those marked with a _are posted on the permanent website: http://www.mel.nist.gov/div821/webdocs-14/events/cri-nano-workshop-05-2008/NIST-Nanomanufacturing-Workshop-Presentation-Schedule.pdf.

Introductions, antitrust: Anne Chaka

- ***Welcoming remarks:*** R. Kayser (NIST Chief Scientist)
- ***Purpose of workshop, deliverables, workshop logistics, facilitated breakout sessions orientation:*** Anne Chaka (NIST)
- ***Working with NIST:*** Terry Lynch (NIST Office of Technology Partnerships)
- ***Overview of outcomes and follow-up from recent relevant Nanotechnology conferences and workshops:*** Mike Postek (NIST)
- ***Nanomanufacturing Overview:*** Mohan "Mano" Manoharan (Surface and Coatings Technology Laboratory, GE Research)
- ***Forest Products:*** Steve Masia (SAPPI Fine Paper)

Automotive: Xingcheng Xiao (Senior Researcher, General Motors R&D)

- ***Pharmaceuticals and Medical Devices:*** Gary Fletcher, (Advanced Technology Leader/ R&D, Becton Dickinson)
- ***Chemicals and Semi-Conductors:*** Mike Garner (Intel)
- ***Aerospace:*** Keith Dan Humfield (Boeing Phantom Works)
- ***Food Science:*** Bernadene Magnuson (Senior Scientific & Regulatory Consultant, Cantox Health Sciences International)
- ***Occupational Safety and Health Issues and Nanotechnology:*** Paul Schulte (Director of the Education and Information Division at NIOSH, and coordinator of the agency's Nanotechnology Research Program)
- ***Challenges in the Synthesis of Nanomaterials for a Wide Variety of Applications:*** Mohamad Al-Sheikhly, Department of Materials Science and Engineering, University of Maryland, College Park

  Nanotechnology Commercialization and Nanomanufacturing: Evgenia Pekarskaya (Lux Associates)

  International Standards: Impact on Nanomanufacturing: Angela Hight-Walker (NIST)
- ***NIST Combinatorial Methods Center:*** Kathryn Beers (NIST)
- ***Nanomanufacturing Overview:*** Mark Tuominen (University of Massachusetts-Amherst) ***Separations/Fractionation:*** Wyatt Vreeland (Biochemical Science Division, NIST)

  Composites: Jeff Gilman (NIST), Huber

  Surfaces/Interfaces: Linda Schadler (RPI)
- ***The United States Measurement System and Measurement Needs for Nanotechnology and NanoEH&S:*** Clare Allocca (NIST)

  NIST Center for Nanoscale Science and Technology: Lloyd Whitman (NIST)
- ***NIST Technology Innovation Program***:Margaret Phillips (NIST)
- ***The Way Forward: Potential interaction models for collaboration and cross-fertilization:*** Anne Chaka (NIST) and Mike Garner (Intel)

APPENDIX D. ACRONYMS

AF&PA	American Forest and Paper Association
AFM	atomic force microscopy
ANI	Applied Nanomat, Inc.
CD	chromatic dispersion
DLS	dynamic light scattering
DNA	deoxyribonucleic acid
EELS	electron energy-loss spectrometer
EH&S	Environmental, Health, and Safety Issues
FFF	field flow fractionation
FIB	focused ion bean
FPLC	fast protein liquid chromatography
GPC	gel permeation chromatography

IEC	ion exchange chromatography
MVD®	Molecular Vapor Deposition
MIT	Massachusetts Institute of Technology
NCN	Network for Computational Nanotechnology
NIH	National Institutes of Health
NIST	National Institute of Standards and Technology
nm	Nanometer
NNI	National Nanotechnology Initiative
NP-HPLC	normal phase-high performance liquid chromatography
NSF	National Science Foundation
NSOM	near-field scanning optical microscopy
PV	photovoltaic
QC/QA	quality control/quality assurance
R&D	research and development
RD&D	research, development, and deployment
RF	radio frequency
RP-HPLC	reverse phase-high performance liquid chromatography
SEC	size-exclusion permeation chromatography
SEM	scanning electron microscopy
SPMs	scanning probe microscopes
STEM	scanning transmission electron microscope
STM	proximal probe microscopies such as scanning tunneling microscopy
TEM	transmission electron microscopy
USFS	U.S. Forest Service
XPS	X-ray photoelectron spectroscopy
ZnO	zinc oxide

End Notes

[1] M. A. Meyers et. al. Progress in Materials Science, **53**, (2008),1-206
[2] "The emerging nanomedicine landscape," V. Wagner et al., Nature Biotechnology 24 (2006) 1211-1217.
[3] "The present and future of nanotechnology in human health care," S.K. Sahoo et al., Nanomed. 3 (2007) 20-31.
[4] "Drug Delivery Systems: Entering the Mainstream," T.M. Allen & P.R. Cullis, Science 303 (2004) 1818-1822.
[5] "Multifunctional nanoparticles – properties and prospects for their use in human medicine," Trends in Biotechnology 26 (2008) 425-433.
[6] "Multimodal optical sensing and analyte specificity using single-walled carbon nanotubes," D.A. Heller et al., Nature Nanotechnology, Published online 14 December 2008.
[7] "Multiplex protein assays based on real-time magnetic nanotagsensing," S.J.Osterfeld et al., PNAS Early Edition December 2008, doi/10.1073/pnas.0810822105.
[8] "Multifunctional nanoparticles – properties and prospects for their use in human medicine," Trends in Biotechnology Vol.26 No.8, 425-433, (2008)
[9] Instrumentation and Metrology in Nanotechnology. National Nanotechnology Initiative Grand Challenge Report. 2006. http://www.nano.gov/NNI_Instrumentation_Metrology_rpt.pdf
[10] Global Industry Analysts, Inc., http://www.strategyr.com/Nanocomposites_Market_Report.asp.
[11] Source: http://www.nanowerk.com/nanotechnology/companies/Hewlett-Packard_Development_Company.html
[12] Source: http://www.appliednanomat.com
[13] Source: http://www.nanowerk.com/nanotechnology/companies/XG_Sciences.html
[14] Source: http://www.udri.udayton.edu/NR/exeres/12BE8330-16AD-4FF4-914A-0CBB9E334E7D.htm
[15] Source: http://www.clarycon.com/plasmacoupling2.html
[16] Source: http://cint.lanl.gov/discovery_platform.html

[17] Source: http://www.appliedmst.com/

[18] Source: http://physics.ucsd.edu/~drs/left_home.htm

[19] Source: http://www.mysticmd.com/solutions.html

[20] Source: http://www.nanowerk.com/nanotechnology/companies/Y-Carbon.html

[21] Environmental, Health and Safety Research Needs for Engineered Nanoscale Materials. National Nanotechnology Initiative. 2006. http://www.nano.gov/NNI_EHS_research_needs.pdf

[22] Toxicology Data Network, U.S. National Library of Medicine. http://toxnet.nlm.nih.gov/

[23] The degree to which a product can be efficiently, cost-effectively and accurately produced using modern manufacturing methods; sometimes achieved through "design for manufacturability."

In: Nanoscience, Nanomanufacturing and Nanoelectronics ISBN: 978-1-61761-816-1
Editor: Joseph G. Devin

Chapter 3

COMPETING FOR THE FUTURE: UNLOCKING SMALL WONDERS FOR TOMORROW'S NANOELECTRONIC USES

National Institute of Standards and Technology

ACKNOWLEDGMENT

The Advanced Technology Program would like to acknowledge the contributions and cooperation of the award recipients highlighted within this chapter, and the following ATP project managers and staff who contributed substantially to the report's completion:

Editors:	Michael Schen, and Lorel Wisniewski
Contributors:	Frank Barros, Carlos Grinspon, Thomas Lettieri, Purabi Mazumdar, and Eric Samuleson
Cover Design/Publishing:	Jo Ann Brooks *Administrative:* Michelle Beddow

> *"Nanotechnology touches upon a broad array of disciplines, including chemistry, biology, physics, computational science, and engineering. Like information technology, nanotechnology has the potential to impact virtually every industry, from aerospace and energy to healthcare and agriculture."*
>
> The National Nanotechnology Initiative at Five Years: Assessment and Recommendations of the National Nanotechnology Advisory Panel
> May 2005

It has been suggested that by 2025 at least half of the companies comprising today's Dow Jones Index will be replaced by nanotechnology companies. In a recent report, Lux Research forecast the worldwide market for nanotech products and services to reach $2.6 trillion by 2014. If this happens, it will represent approximately 15% of global manufacturing output.[1]

Nanoelectronics is defined as advanced electronic and optical systems that depend upon innovations in nanotechnology. The field of nanoelectronics embraces a range of technologies, all ultimately for the purpose of creating and improving the functionality, cost, and performance of tomorrow's electronic systems. This chapter examines the historical role the Advanced Technology Program (ATP) has played in fueling innovation and American competitiveness in areas of nanotechnology that will lead to totally new-to-the-world electronic and optical products.

NIST, ATP and Nanotechnology

For more than a century, the National Institute of Standards and Technology (NIST) has helped to lay the foundation for the innovation, economic growth, and quality of life that Americans have come to expect. NIST technology, measurements, and standards help U.S. industry invent and manufacture superior products reliably, provide critical services, ensure a fair marketplace for consumers and businesses, and promote acceptance of U.S. products in foreign markets.

Researchers in NIST's seven major laboratories are developing measurements, standards, and data crucial to private industry's development of products for nanotechnology markets. NIST's work also aids federal agencies' efforts to exploit nanotechnology to further their missions, such as national security and environmental protection. In addition, NIST has launched the new state-of-the-art Center for Nanoscale Science and Technology to help U.S. companies, universities, and government agencies overcome obstacles to cost-effective manufacturing of nanoscale products.[2] NIST's accomplishments in nanotechnology are captured in a recent NIST Special Publication.[3]

> The nanotechnology projects funded by ATP are poised to impact the way in which tomorrow's electronic and optical products are conceived, perform, and ultimately deliver economic security, quality of life, and social benefits to American citizens.

The Advanced Technology Program (ATP), one of the three external programs within NIST, has contributed to developments in nanoelectronics since 1999 when the first nanotechnology awards began. ATP has the mission of helping U.S. companies bridge the gap between the research lab and the marketplace by encouraging them to pursue the development of high-risk enabling technologies that otherwise would not be developed or that would be developed too slowly to take advantage of market opportunities. ATP's early stage partnerships and investments with the private sector accelerate the development of innovative technologies deemed too technically risky or too far from commercial viability to be considered by private sector funding sources. ATP-funded projects promise significant commercial payoffs and widespread economic benefits for the nation in the longer term. Through a competitive peer review process, ATP has issued awards based on rigorous technical and economic criteria:

- Scientific and technological merit: innovation, high technical risk and feasibility, and quality of R&D plan, and
- Potential for broad-based economic benefit: national economic benefits, need for ATP funding, and pathway to economic benefits.

ATP has funded **25 nanotechnology awards** across nine states for nanoelectronic applications. These awards represent **$224.1 million of total industry-based R&D,** of which $114.6 million has come from ATP and $109.5 million as cost share from industry. ATP's first awards in nanotechnology became active in Fiscal Year (FY) 1999 and addressed technology barriers in semiconductor manufacturing and energy. Since then, the number of nanotechnology awards has steadily grown as opportunities emerge. **Figure 1** presents the profile of ATP nanotechnology awards for nanoelectronics underway during any one year, from FY 1999 to the present. For example, 18 of the 25 awards were active and underway in FY 2005.

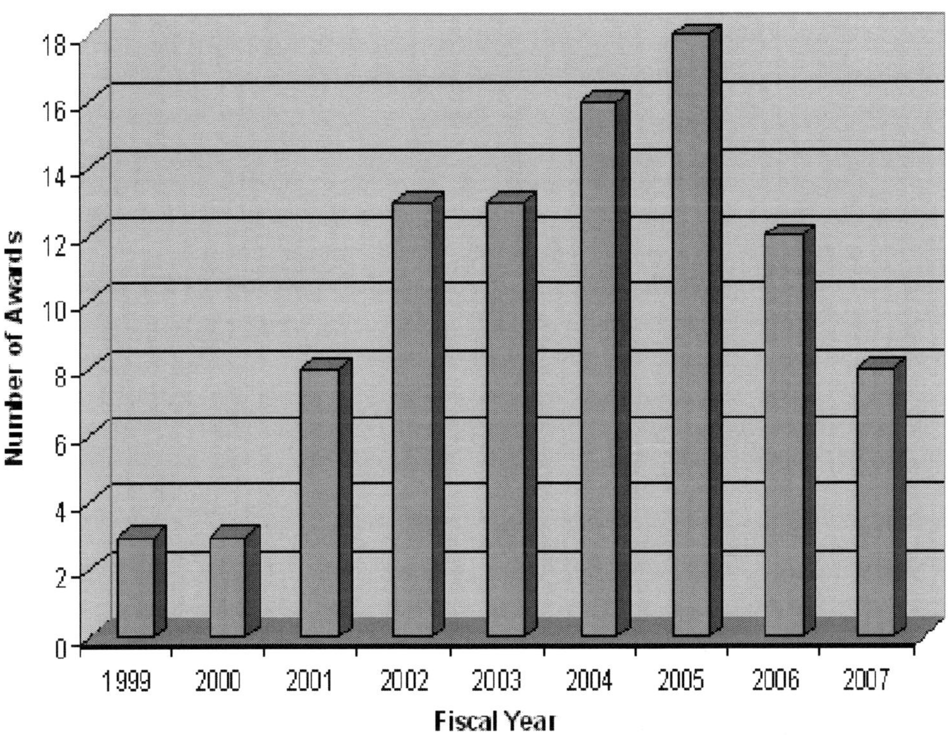

Figure 1. Yearly Representation of Active ATP Nanotechnology Awards for Nanoelectronic Uses

Table 1. ATP Nanotechnology for Nanoelectronics Awards and Funding ($ millions)

Total R&D Funded	**$224.1**	**Total Participants** [5]	**90**
ATP share	$114.6	Small Company	43
Industry share	$109.5	Medium Company	7
		Large Company	16
Total Awards	**25**	Not-for-Profit Organization	1
Single Applicant	13	University	22
Joint Venture	12	Federal Laboratory	1
Single Applicant by Lead	**13 awards**	**Joint Venture by Lead**	**12 awards**
Small Company	13	Small Company	5
Medium Company	0	Medium Company	2
Large Company	0	Large Company	4
		Not-for-Profit Organization	1

Table 1 summarizes ATP's nanotechnology funding for projects that impact nanoelectronics. It shows the distribution of awards among award type, participant type and size[4], and presents project lead information by company size. Of the 25 awards funded, 13 have been single applicants and 12 as JVs. Within these awarded projects, a total of 90 entities have participated as recipients and subcontractors. Of these, nearly 80% of all participants have been small entrepreneurs, mid-sized companies, or universities. Furthermore, *all* single applicant awards and *over half* of the joint-venture (JV) awards for nanoelectronics have been led by small- or medium-sized companies. This large percentage involvement by small- to medium-sized businesses, and universities, speaks to the newness and opportunities of nanotechnology, and the role these enterprises play in creating game-changing, disruptive technologies. *Appendix A* shows all companies, universities, or other entities that have participated in ATP awards in nanotechnology for tomorrow's electronics.

In the nanotechnology awards reported, universities as subcontractors or JV partners are playing an essential and expansive role in translating laboratory research into viable commercial technologies. Universities participate, on average, in nearly each award ATP has funded in this area. Also, ATP JVs are being effectively used to foster partnerships between universities as scientific experts, small entrepreneurs as suppliers, and large firms as developers of complex technological systems. ATP nanotechnology JV awards for nanoelectronics represent the cutting edge in technology innovation and in the business model used by international systems developers to remain competitive within the global marketplace. ATP's awards in this area are helping position American small businesses to become best-of-the-breed global suppliers.

Table 2 shows the funding distribution among states, based on the location of the lead company for the award. The funding distribution shown does not reflect the final funding received by each state because projects distribute funds to several partners and subcontractors that may be located in other states, including those not shown here. *Appendix A* does identify states that are associated with each ATP award participant and the nature of their participation.

Table 2. ATP Nanotechnology for Nanoelectronics Awards and Funding Among States

State	AWARDS			FUNDING*		
	Total	Single Company	Joint Venture	Total R&D, $M	ATP Share, $M	Company Share, $M
California	8	5	3	$62.69	$32.63	$30.06
Connecticut	1	1	0	$0.90	$0.44	$0.46
Kansas	1	1	0	$2.31	$2.00	$0.31
Maryland	1	1	0	$3.38	$1.97	$1.41
Massachusetts	1	1	0	$3.41	$2.00	$1.41
Michigan	2	0	2	$30.99	$15.11	$15.88
New Mexico	1	1	0	$3.84	$2.00	$1.84
New York	3	1	2	$16.90	$8.22	$8.68
North Carolina	2	1	1	$9.35	$5.38	$3.97
Texas	5	1	4	$90.31	$44.80	$45.51

* Funding amounts are a reflection of the state of residence of each award lead company.

EXPANDING FRONTIERS AND DELIVERING BENEFITS: INNOVATIONS IN NANOTECHNOLOGY

> The New York State Office of Science Technology and Academic Research (NYSTAR) reported earlier this year (2006) that the State University at Albany's Center for Advanced Technology in Nanomaterials and Nanoelectronics has generated $1.1 billion in economic impact against a New York State investment of $7 million.
>
> *Albany Business Review*, February 27, 2006

Today, across the globe, a race is underway as scientists and engineers seek to unlock the secrets and possibilities of nanotechnology. [6] The ability to understand and control matter at dimensions from 1 to 100 nanometers, where unique phenomena arise, will unlock innovation and enable novel applications. The nanotechnology innovations made possible by ATP funding will aid America's economic security in tomorrow's information, health, and energy economies.

Table 3. Early-on Application Areas of ATP Funded Projects in Nanoelectronics

Application Area	Number of Awards	Funding		
		Total R&D, $M	ATP Share, $M	Company Share, $M
Alternative Energy	5	$18.38	$9.93	$8.44
Bioelectronics	3	$16.74	$9.60	$7.14
Materials Discovery	2	$16.56	$8.56	$8.00
Mixed-function Electronic Systems	7	$73.32	$36.21	$37.11
Semiconductor Nanomanufacturing	8	$99.09	$50.26	$48.83

Table 3 shows five application areas that emerge as the most likely early-on beneficiaries of the nanotechnology innovations that have been developed by ATP funded project teams. Because of the enabling nature of the awards ATP funds, spill-over benefits to other application areas are expected to accrue as the technology is successfully diffused within the marketplace. Thus, application areas beyond those identified below are likely to emerge from the awards ATP has funded in these areas.

Nanoelectronics, like nanotechnology, is already positioned to be a major engine of change that drives economies well beyond the 21st Century. Many of ATP's nanotechnology awards for nanoelectronics directly involve innovations in metrology, materials, manufacturing, or device technologies that directly impact semiconductor nanomanufacturing and electronic systems application areas.

- ***Semiconductor nanomanufacturing*** investments set the stage for tomorrow's integrated circuits, the infrastructure upon which they depend, and the nanoelectronics products that they enable.
- ***Mixed-function electronic systems*** benefit from nanotechnology innovations in magnetic, optical, or electronic technologies. When merged with advanced semiconductor electronics, giving rise to broader utility of nanoelectronics in tomorrow's economy takes place.

Nanoelectronics offer solutions to the demands for ever increasing data rates. It enables a shift to optical communications within and between IC chips. New low-cost ultra-high bandwidth optical communications, made possible by nanomanufacturing photonic crystals, will greatly improve the performance of computers and the internet. Other advancements include nanomaterials for solid-state lighting; precise deposition reactors for ultra-high density magnetic memory; and precisely controlled growth of nanomaterials for medical imaging systems, lamps and displays. These innovations will be possible because of ATP funding that benefits mixed-function electronic systems.

> New nanoscale devices and manufacturing processes are allowing manufacturers to utilize spintronics-based memory in conventional integrated circuits, allowing the increased use of magnetic memory for non-volatile, low power, and fast computer memory.

Three other nanoelectronics application areas expected to find early benefits from ATP funded projects include bioelectronics, fundamental materials discovery, and alternative energy generation.

- ***Bioelectronics,*** an interdisciplinary research field that has recently emerged, will impact a range of industries. It seeks to exploit the growing technical ability to integrate biomolecules with electronics to develop a broad range of functional devices.[7] Bioelectronics includes elements of chemistry, biology, physics, electronics, nanotechnology, and materials science.

- ***Materials discovery*** is the process of creating wholly new materials, including electronic materials or nanomaterials, as a result of nano-scale metrology innovations being coupled with advanced electronic systems.
- ***Alternative energy,*** in particular renewable energy such as fuel cells, photovoltaics, and energy storage systems, all are poised to benefit from breakthroughs in nano-materials, devices and manufacturing.

To be commercially successful, nanoelectronics, like other applications of nanotechnology, demands a broad range of innovations. The federal government first acted on the potential for the broad-based impact of nanotechnology in FY 2001 through the establishment of the National Nanotechnology Initiative (NNI). The NNI identifies a number of essential investment areas in nanotechnology innovation.[8] Four NNI innovation areas are applicable to the awards ATP funded:

- New nanoscale *devices and systems* built upon diverse nanotechnology innovations with improved performance or new functionality.
- Cutting edge *instrumentation and metrology* capable of resolution, precision and throughputs that match industry's exacting demands.
- New *manufacturing and process technologies* that allow for precisely controlled fabrication of nanometer structures that perform and behave within design tolerances.
- New *nanomaterials* that capture uniquely new properties and functionality as a result of their controlled composition and nanometer-scale structure.

Figure 2 presents a breakdown of ATP's nanotechnology awards with application to nanoelectronics, using the four NNI innovation areas outlined above. New nanomaterials innovations dominate the portfolio, accounting for 36% of ATP's awards impacting nanoelectronics. This observation underscores the enabling nature of new materials innovations that has historically driven new technology innovation.

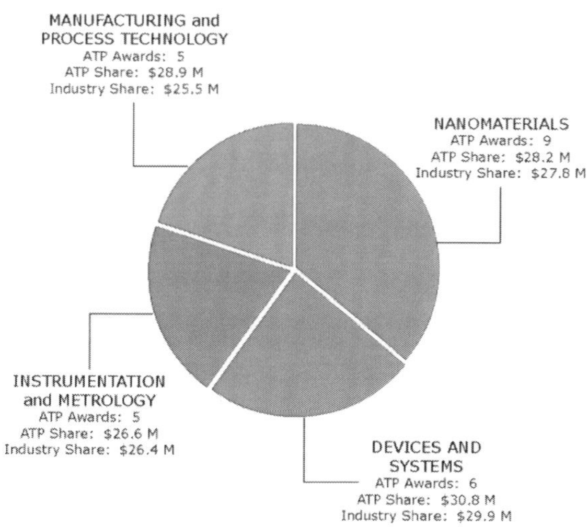

Figure 2. ATP Funded Nanotechnology Innovation for Nanoelectronics Applications

The following sections present examples of ATP-funded projects in each of these four NNI innovation areas, and provide a glimpse into the technologies that have been developed and the potential economic security, quality of life, and societal benefits to America's citizens that are likely as a result of ATP funding. *Appendix B* also presents all 25 ATP nanotechnology awards for nanoelectronics categorized by NNI innovation and early-on application areas.

Devices and Systems

ATP has invested $30.8 million in nanoscale devices and systems R&D, with $29.9 million industry cost share across six projects.

Innovations in nano-scale devices and systems are poised to provide new types of electronic, photonic, and other devices and systems across a wide range of applications. Biotechnology and healthcare, and telecommunications and integrated circuits are a few of the application areas most likely to benefit from innovative nano-scale devices and systems.

An ATP funded award in nanoscale devices and systems that is well positioned to create substantial broad-based benefits for U.S. industry was recently completed by Luxtera, Inc.[9]

Nanophotonic Integrated Circuits for Telecommunications and Computing

Luxtera Inc.

Image: Electron microscope image of a vertical optical coupler within a npIC (image courtesy of Luxtera, Inc.)

Project: Design, fabricate, and demonstrate the performance of nanophotonic circuits integrated on semiconductor wafers using standard semiconductor processing equipment.

Optical switching systems now used to send telephone calls and internet data over fiber-optic communication lines fill large racks and are costly to buy and operate. In its ATP project, Luxtera, Inc. drastically shrank these systems down to just a few microchips by designing, fabricating, and demonstrating the performance of nanophotonic circuits integrated on to semiconductor wafers using standard semiconductor processing equipment.

The nanophotonic Integrated Circuit (npIC) technology represents a new industry segment offering substantial product cost and performance advantages in telecom and microprocessor applications. The Luxtera advance will enable microprocessors to communicate with each other at ultra-high speeds, reducing the need for memory cache and increasing the efficiency of multiprocessor applications. According to a recent article in *Forbes* magazine, "teaching silicon new photonic tricks promises a huge boost in getting data out of a computer." [10] The article notes that Luxtera's growing fan club includes Michael Fister, formerly with Intel and now chief executive of Cadence Design Systems, the leading maker of chip-design software tools. "There's good reason to be intrigued with this technology, and Luxtera is definitely worth watching," Fister states.

> "The potential impact on the industry of combining photonic and electronic elements on a single CMOS die is substantial," said Lawrence Gasman, president of CIR, a leading technology forecasting company. "Many applications, including those in the cost sensitive consumer markets, will benefit from the improvements in cost, power consumption and size."

In an August 2006 press release, Luxtera announced the industry's first single-chip npIC device implemented in a standard CMOS process. The technology integrates high-performance optics and mainstream electronics on a single die, bringing fiber connectivity directly to the chip. According to the company, "Additional digital logic can be integrated into the same chip with optical devices, further reducing overall solution size, power consumption and cost."

Luxtera reports it is currently sampling prototype devices for preliminary testing by strategic development partners, and plans to have a commercial product line based on this underlying technology by early 2007. Initial product offerings are expected to be multi-port transceivers for communications, storage, and computing applications. The company believes CMOS photonics technology will enable the widespread adoption of 10G interconnects, which today are very expensive to deploy, by driving the cost of 10G optical ports to well below $100.

Total economic benefits to original equipment manufacturers, chipmakers, telecom service providers, businesses, and consumers are projected to be substantial.

Further examples of ATP funding of high technical risk R&D in nanoscale devices and systems are shown below.

Nano-Imprint Lithography Infrastructure for Low Cost Replication at the 65 nm Node and Beyond

Participants: Molecular Imprints Inc.; Motorola Inc.-Motorola Labs; Photonics Inc.; KLA-Tencor Corporation; and University of Texas at Austin

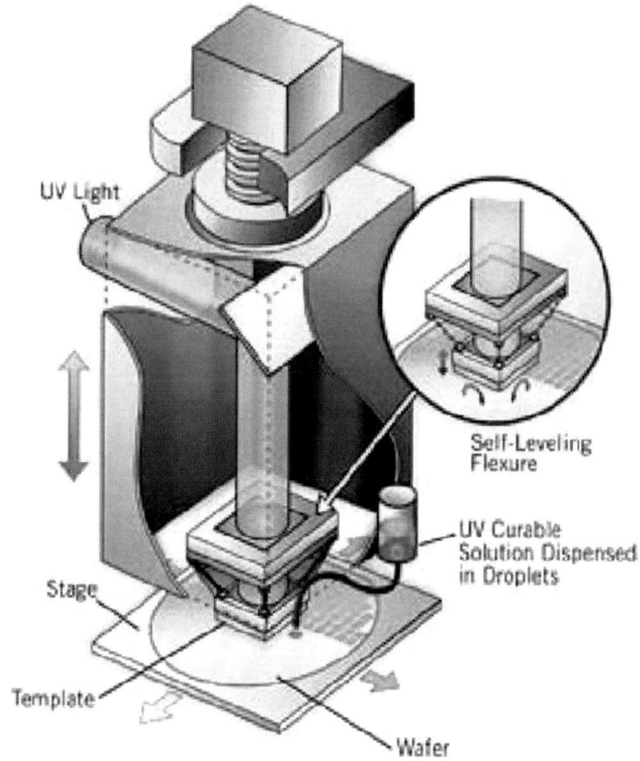

Image: Schematic representation of MII nanoimprint lithography tool (image courtesy of Molecular Imprints, Inc.).

Design and demonstrate technology for step and flash imprint lithography (S-FIL), a novel method of transferring integrated circuit patterns to the surface of a semiconductor wafer by molding of three-dimensional features potentially as small as 20 nanometers or less.

Molecular Imprints Wins EE Times' Ace Award For "Most Promising New Technology", March 15, 2005

Molecular Imprints, Inc. (MII) won the EE Times' Annual Creativity in Electronics (ACE) Award for "Most Promising New Technology", beating out IBM, Intel, and two others for this award. MII's technology, called Step and Flash Imprint Lithography (S-FIL™), was chosen for this award, which goes to a compelling electronic component or enabling technology that is outstanding in its technical design, potential for market impact, and demonstrated leadership in its area.

The Seamless Detection and Treatment of Cancer with Near-Infrared Absorbing Nanoshells

Participant: Nanospectra Biosciences, Inc.

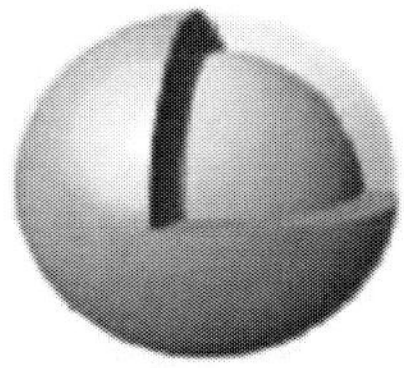

Image: Simulated representation of a near-infrared absorbing nanoshell (courtesy of Nanospectra Biosciences, Inc.).

Develop an integrated approach to the diagnosis and treatment of cancer allowing more accurate detection and contemporaneous, minimally invasive treatment using near-infrared absorbing nanoparticles.

Deposition Source for Producing Super Lattice, Multilayer Thin Films to Enable Perpendicular Magnetic Recording

Participant: Intevac, Inc.

Develop a high-throughput thin-film deposition source capable of rapidly depositing precisely defined multilayer thin films to enable commercial production of magnetic super lattice films for the next generation of high-capacity disk storage devices.

Image: The Intevac 200 Lean (courtesy of Intevac, Inc.

INSTRUMENTATION AND METROLOGY

ATP has invested **$26.6 million** in nanotechnology instrumentation and metrology R&D, with **$26.4 million industry cost share** across **five projects.**

U.S. manufacturers require a wide range of tools for inspecting and measuring their output for quality and performance. Metrology is essential for development and improvement of new materials, processes, and tools for manufacturing; all essential for successful commercialization of nanomanufacturing.

Intevac Delivers a Revolution in Data Storage

In their ATP project, Intevac, Inc. developed a revolutionary magnetic disk sputtering system tool for manufacturing next generation magnetic data storage known as perpendicular magnetic recording. New perpendicular media recording technology delivers smaller form factor magnetic hard disk drives that meet ever-increasing demands of consumer products, and achieve the productivity and costs requirements for future hard disk manufacturing.

The ATP-funded R&D has given rise to a new Intevac tool named *200 Lean*. The *200 Lean* is Intevac's latest-generation disk sputtering system with capabilities to address advanced longitudinal and perpendicular media processing on the same platform.

According to Intevac in their June 12, 2006 press release, the company reports, "we project that Intevac would deliver between 34 and 39 *200 Leans* for revenue during calendar 2006."

"Metrology is a key enabler for all manufacturing and it is especially important to nanotechnology. It has been predicted that within the next 10 years, at least half of the newly designed advanced materials and manufacturing processes will be built at the nanoscale. Measurement science (metrology) and advanced instrumentation are essential for nanomanufacturing"[11]

An ATP funded award in nanotechnology instrumentation and metrology that is well positioned to create substantial broad-based benefits for U.S. industry was completed by Micro Magnetics, Inc. [12]

Photo by Barry Gardner

Spintronics-Based High-Resolution, Non-Invasive, and Ultrafast Metrology for the Semiconductor Industry

Participant: Micro Magnetics, Inc.

Project: Develop a nanoscale magnetic tunnel junction current-sensing system for integrated circuit inspection that will help maintain the U.S. lead in semiconductors by providing vastly improved metrology for in-process inspection.

Micro Magnetics develops, produces and markets novel technologies employing advances in magnetic sensing technology based on spintronics–a revolutionary new field created by the convergence of magnetics, electronics and nanotechnology. These technologies aim to provide the semiconductor and manufacturing industries with more efficient, cheaper products based on the introduction of technologies enabled by high-performance miniaturized magnetic sensors and devices.

Recently, Micro Magnetics announced a series of new product offerings based on the ATP technology, featuring low-field sensors with 180% magnetoresistance. These sensors feature magnetic tunnel junction devices that incorporate cutting-edge magnesium oxide barrier technology. This capability represents a sixfold increase in sensitivity and voltage response from their current SpinTJ products and will enable a new generation of low-field sensing applications.

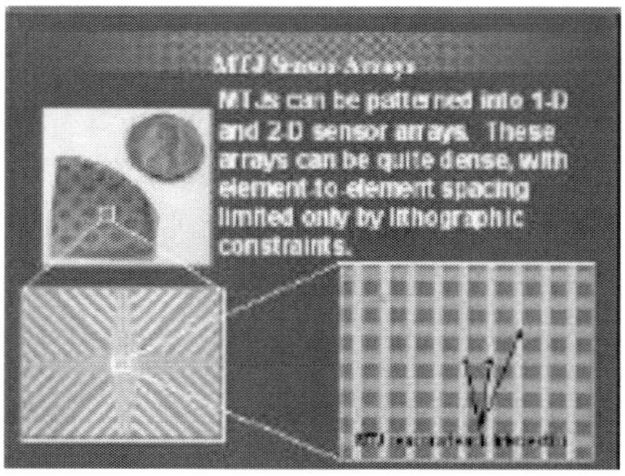

Image: Micro Magnetics MTJ sensor array (courtesy of Micro Magnetics, Inc.)

Micro Magnetics is the first company worldwide to offer low-field sensors with magnetoresis-tance values in excess of 100%, an important milestone in magnetic sensing. The smallest dimension of the MTJ sensor is only about 10 nanometers, with field sensitivity of 5 nanotesla, active areas as small as 1x2 micrometers, and an operating temperature as high as 100° C.

The ATP funded R&D is being driven by opportunities in semiconductor nanomanufacturing, specifically semiconductor fault isolation and failure analysis.

Spill-over applications of the ATP funded MTJ sensor technology are also being pursued by Micro Magnetics, Inc. in the following areas:

- ***Bio-sensing and Biomagnetic Applications:*** Micro Magnetics' micrometer-size magnetic tunnel junction sensors have demonstrated the capability to detect single magnetic nanoparticles, which makes them ideal for use in the emerging applications area of "lab-on-a-chip" detectors.
- ***Detection of Magnetic Media and Inks:*** The magnetic sensors will be utilized in many of today's most vibrant industrial segments such as Hard-disk drives, Digital and analog video, audio tapes, flight data recorders, and floppy disks.

Micro Magnetics' Circuit Scan 1000 is a magnetic imaging tool which uses the tiny magnetic fields emitted by all of the currents inside any chip to understand the inner workings of the circuitry– without the need to process or even touch the device at all. Using mathematical algorithms which convert the magnetic field data into a map of current density, this system allows the user to see a map of all of the current flowing on every level of the device, down to the single microamp level. Because current is the lifeblood of an integrated circuit, this map can be invaluable to engineers and technicians attempting to understand the nature of the problem.

Further examples of ATP funding of high technical risk R&D in instrumentation and metrology are shown below.

Advanced Technology for Non-destructive, Localized, Dielectric Metrology of Future Generation Integrated Circuits
Participant: Neocera, Inc.

Develop technology for making non-destructive, quantitative measurements of dielectric constants at length scales (about 100 nanometers) and microwave frequencies (1-10 gigahertz) appropriate to ICs. Potential impacts include:

- Accelerate process development and enhance manufacturing yield for ICs,
- Decrease time to market of future ICs, and
- Create cost savings for IC manufacturing.

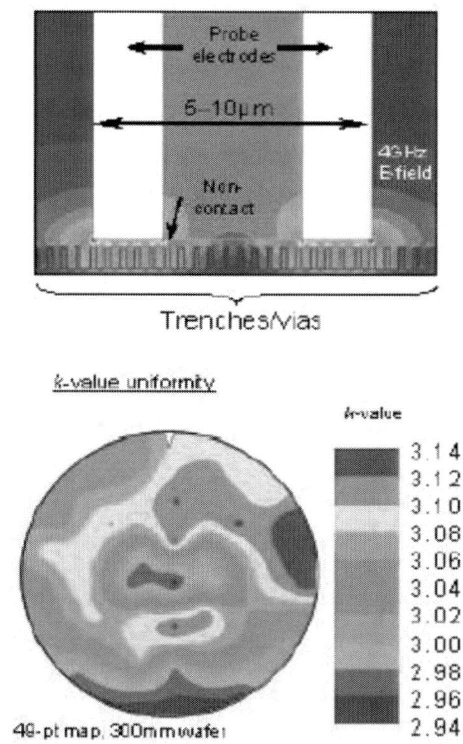

Probe simulation and wafer level k-value uniformity profile (images courtesy of Neocera, Inc.)

High-Speed AFM-Based Platform for Quantitative Nanomechanical Measurements
Participants: Dow Chemical Company; and Veeco Instruments, Inc.

Develop an atomic force microscopy based platform for high-speed, high-bandwidth quantitative nanomechanical measurements which provides structure-property understanding at the nanoscale in order to accelerate discovery, design, and commercialization of nanomaterials.

Potential Impacts of this technology include:

- Unprecedented mechanical property resolution at interfaces for new product development and predictive material models;
- Development of nanocomposites that can save fuel costs by lowering the overall weight of automobiles;
- Reduce yield loss in sub-65 nm semiconductor manufacturing; and
- Increase the speed of critical characterization for nanoscience research.

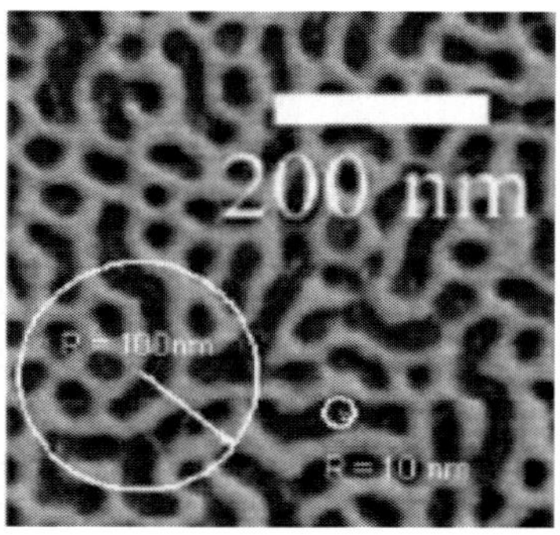

Image: The dimensions of a current nanoindentation tip (R = 100 nm) and proposed tip (R=10 nm) superimposed on a developmental copolymer film (imaged courtesy of Dow Chemical Co.)

MANUFACTURING AND PROCESS TECHNOLOGIES

ATP has invested **$28.9 million** in nanotechnology manufacturing and process technologies R&D, with **$25.5 million industry cost share** across **five projects**.

Manufacturing at the nanoscale level can be achieved by either scaling down current process technologies or by molecular level self-assembly process technologies. R&D for implementing these two approaches include pattern generation in 2-dimensional and 3-dimensional structures, atomically precise manufacturing processes, tools for defect identification and reduction, and new processes to fabricate nanoscale structures. New manufacturing technologies should also be friendly to the environment, i.e. reduce the use of hazardous chemicals and emission of pollutants, and conserve natural resources.

An ATP funded award in nanotechnology manufacturing and process technology that is well positioned to create substantial broad-based benefits for U.S. industry was recently completed by Uncopiers, Inc. [13]

ACIM "Point*Suns": Concentrating Energy through Silent Sound and Clean Water
Participant: Uncopiers, Inc.

Project: Design and build an energy-efficient, chemical-free nanoparticle detector and wafer cleaner to enable the semiconductor industry to clean and inspect next-generation wafers reliably and profitably. Locating a nanoparticle on a wafer is equivalent to finding a specific grain of sand on a baseball field.

Uncopiers, Inc., a start-up company located in Manhattan, Kansas, is developing a new technology, called PointSuns™, ideally suited for nanotechnology pursuits. It is a unique way of concentrating energy at nanometer length scales without any collateral energy dispersal, dissipation, or damage. Using only silent sound and clean water, PointSuns™ are created through controlled acoustic nano-cavitation. They are capable of pin-pointedly interacting up to several hundred MeV of energy per cubic-nanometer of matter. This new capability allows one to accomplish tasks sometimes considered beyond the reach of prevailing technological practices.

Image: Nanocavitation bubbles at work using silent sound and clean water: here it is shown removing ink from paper (courtesy of Uncopiers, Inc.)

The ATP funded R&D is being used to overcome critical issues in semiconductor nanomanufacturing:

- ***Semiconductor wafer cleaning:*** Uncopiers has developed an entirely chemical free method of removing sub-micrometer particulates from semiconductor wafers. Only water and silent sound are used to dislodge attached nanoparticles from the wafer surface, rendering the wafer clean for further processing. The water effluent itself can be refiltered, reprocessed and reused, thus conserving a precious resource. Because no chemicals are used, the cleaned wafer can be rinsed and dried in the same setting on a common footprint. Wafer cleaning is the most frequently used process step in IC manufacturing.

- ***Detection and elimination of nanoparticles in liquids:*** Controlled cavitation can be used to detect the presence of nanoparticles in liquids. Presently available methods of particle detection fail to detect sub-100nm individual particles in liquids. Detection of nanoparticles in liquids paves the way for ultrapurification of liquids critical in IC manufacturing.

Spill-over applications of the ATP funded PointSuns™ technology are also being pursued by Uncopiers, Inc. in the following areas:

- ***Adhesion testing of thin films:*** In another application, PointSuns™ can be used to characterize and fashion thin films on substrates. Adhesion integrity of monolayer thin coatings can be conclusively and quantitatively inferred using this new technology.
- ***Removal of ink from paper in non-destructive way:*** The ATP PointSuns™ technology can be used to completely remove the photocopier ink from paper without hurting the underlying paper in an effort to reuse-recycle paper.

Further examples of ATP funding of high technical risk R&D in manufacturing and process technology are shown below.

Rapid and Comprehensive Development of Advanced Dielectric Materials for Wireless Applications

Participant: Intematix Corporation

Develop novel advanced dielectric materials using high-throughput synthesis and diagnostic tools to enable path-breaking improvements in the next generation of wireless telecommunications components.

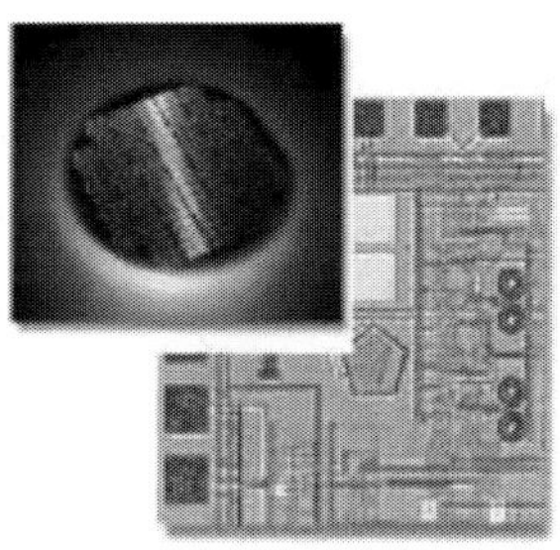

Image: Typical semiconductor wafer and circuit pattern.

Small Times, 2005

Intermatix was named one of five finalists for **2005 Best Product of the Year** by ***Small Times*** magazine. Using its ultra-rapid materials development and proprietary tools and processes, Intermatix discovers materials in a fraction of the time typical for conventional methods.

Assemblers for Nanotechnology Applications and Manufacturing: Enabling the Nanotechnology Era

Participants: Zyvex Corporation; and Honeywell International, Inc.

Development of low-cost, computer controlled, microscale assemblers that operates in parallel to assemble three-dimensional microscale components, with extension of this technology to nanoscale assemblers for the commercialization of nanotechnology.

Image: Optical image of an ion source for mass spectrometers built with automated microassembly (courtesy of Zyvex Corp.)

Building upon the ATP funded technology, a number of application areas are being pursued by Zyvex Corp.

- ***Microsensors:*** Recent commercialization efforts have focused on miniature optical systems (electron, ion and photon) for micro-instrumentation and sensing applications such as miniature mass spectrometers (MMS) and miniature scanning electron microscopes. The automated microassembly technology, developed under the ATP project, has enabled the creation of these types of systems in an economically viable way.
- ***Biological Tissue Manipulation and Testing:*** The scaled MEMS grippers have potential application in biological tissue manipulation and testing (e.g. collagen fibers) in biomedical research.
- ***Micromanipulation:*** Microgrippers will have applications in TEM sample preparation for semiconductor companies, and for manipulation of components in disk drive assembly.
- ***Assemblers for nanomanufacturing***
- ***Atomically precise manufacturing***

Nanomaterials

ATP has invested **$28.2 million** in nanomaterials R&D, with **$27.8 million industry cost share** across **nine projects**.

At the nanoscale, the physical, chemical, electrical and biological properties of materials differ in often unexpected and sometimes valuable ways from the behavior of bulk matter. Assembling materials with molecular precision create material ensembles with new properties and enable new applications. As a result of imaging, measuring, modeling, and manipulating matter at the nano-scale regime, scientific breakthroughs are being realized that will lead to entirely new approaches and solutions to problems in virtually all areas of technology and the economy.

An example of an ATP funded joint-venture award in nanomaterials that is well positioned to create substantial broad-based benefits for U.S. economy is being led by Cree, Inc. [14]

Low Cost, High Efficiency LED Lamp

Participants: Cree, Inc., and Nanocrystal Lighting Corp.

Project: Demonstrate a white light-emitting diode (LED) lamp package with an integrated chip approach that would more than quadruple the brightness and double the efficiency of existing LED systems and significantly reduce the cost per lumen.

Current lighting technologies are very inefficient, wasting energy that could translate to reduction in spending by individuals and businesses. Specifically, incandescent and halogen lamps consume 42 percent of the total energy used for lighting in the United States while producing only 16 percent of the nation's electrically powered lighting.

A technology that will radically improve this situation is solid-state light-emitting diodes (LEDs) – very small light bulbs illuminated by the movement of electrons in semiconductor materials. LEDs are compact and durable and are often used in traffic lights, small liquid crystal displays, and indicators in cars and electronics.

But while the energy efficiency of white LEDs today exceeds that of incandescent lamps, the overall brightness and cost of high flux LED lamps need to be improved before the technology can penetrate the general illumination market. Cree, Inc., and its joint venture partner Nanocrystal Lighting Corporation, are demonstrating novel integrated white LED lamps that are significantly brighter and more efficient than existing LED lamps, as well as lower in cost per lumen of light output. The companies combined high-efficiency blue LEDs with phosphor and other nanomaterials that convert blue light to white light and significantly reduce energy consumption.

The U.S. Department of Energy estimates, widespread adoption of LED-based lighting would greatly reduce U.S. energy consumption, potentially saving around $10 billion annually by 2011, while also reducing annual carbon emissions by 5-10 million metric tons. The project is expected to generate jobs in the nanotechnology and semiconductor industries in the United States.

In August 2006, Cree announced the release of a new LED power chip that incorporates some of the key innovations developed under the ATP-funded research. The new LEDs are twice the brightness of Cree's current power chips, and are designed for general lighting applications, such as home and office lighting, auto headlamps, streetlights, and garage and warehouse low bay lighting.

Significant breakthroughs are still needed in both materials and packaging fabrication technology to improve light extraction while controlling costs and improving the stability of novel encapsulation materials. The ATP JV brought together two companies with complementary expertise: Cree, a leading maker of LEDs, and Nanocrystal Lighting, a startup formed to commercialize new nano-photonic materials.

Although the current progress in the Power LED technology base is significant, not all the technical goals of the ongoing ATP project have been achieved given that the award is still underway. These potential new breakthroughs could further reduce the future price of white solid-state lighting by up to 50 percent, which together with higher efficiency and longer lifetimes, could lead to significantly greater reach in home and medical lighting applications, backlighting for computers and television, and mobile appliances and projectors, among others. The worldwide market for solid-state lighting could be worth as much as $40 billion.

Further examples of ATP funding of high technical risk R&D in nanoelectronic materials are shown below.

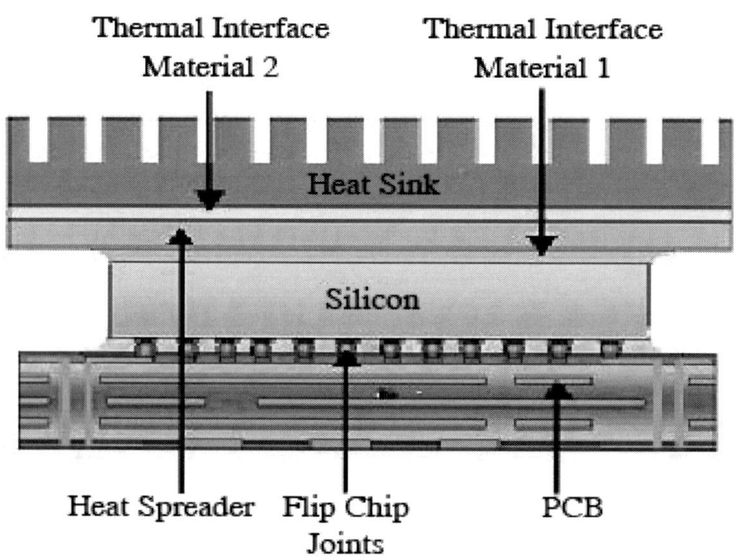

Image: Schematic representation of the use of thermal interface materials in a flip-chip application (courtesy of General Electric Co.)

Nanoengineered Thermal Interfaces Enabling Next Generation Microelectronics

Participants: General Electric Company, University of Binghamton (Binghamton, NY), and Cabot Corporation (formerly Superior MicroPowders, LLC)

Develop and demonstrate the performance of novel materials, for use as interfaces between computer chips and heat sinks, that conduct heat 10 times better than today's interface materials

"Free Standing", Single-Wall Carbon-Nanotube Fuel Cell Electrode

Participants: Carbon Nanotechnologies, Inc.; Motorola; and Johnson Matthey Fuel Cells, Inc.

Developing novel, free-standing single-wall carbon nanotube electrode assemblies containing an immobilized noble-metal catalyst for PEM fuel cells.

Major Patent Milestone

In early 2005, ATP recipient Carbon Nanotechnologies, Inc. (CNI), Houston, TX, announced it had reached a major milestone with 30 issued or allowed patents related to the innovative use of small diameter carbon nanotubes. CNI is a manufacturer and producer of single-wall and other small-diameter carbon nanotubes, thought by many to represent a transforming technology to produce and deliver superior products.

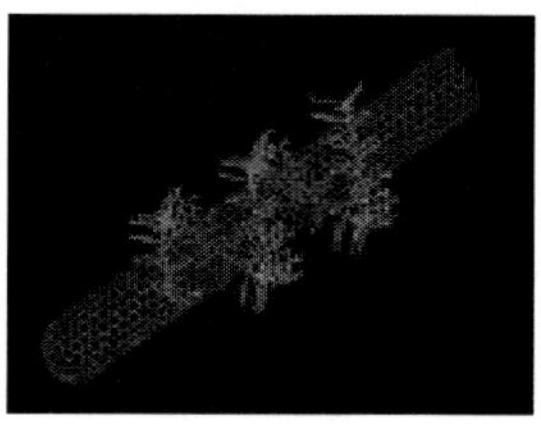

Image: Model-derived representation of a single-wall carbon-nanotube (courtesy of Carbon Nanotechnologies, Inc.)

SUMMARY

Funding for science and technology development programs is typically directed to universities for basic or applied research, or companies for low technical risk product development. Federal funding for basic research is addressing fundamental barriers in nanotechnology and opening up many new opportunities for nanotechnology innovation by U.S. companies. Significant challenges remain in funding translational research in nanotechnology to speed the development of path-breaking technologies that have the potential to create substantial economic benefits.

A funding gap exists for early stage technology development where technical risks remain high and the timeline to success is long. Investments made by the Advanced Technology Program in nanotechnology have helped to bridge this funding gap, facilitated the development of nanoelectronics by U.S. companies and their university partners, and brought to the global marketplace early-on benefits in a number of nanoelectronics application areas.

The commercialization of new, innovative ideas in nanotechnology, made possible by ATP funding, is expected to expand in the years to come. As this occurs, tremendous national benefits, in the form of economic security, quality of life, and societal benefits to American citizens, are expected.

Appendix A. Advanced Technology Program Participants in ATP Nanotechnology Awards for Nanoelectronic Applications

Organization	Project Number	Project Title	Participation Type	State	Start Date	End Date
Small-size Companies						
Abeam Technologies	2004-1 E-6814	Development and Demonstration of a Multiple, High- Current-Density Shaped E-Beam Column With Independent Vector Beam Placement	Subcontractor	California	10/1/2004	9/30/2006
Accurel, Inc.	2004-1 E-6814	Development and Demonstration of a Multiple, HighCurrent-Density Shaped E-Beam Column With Independent Vector Beam Placement	Subcontractor	California	10/1/2004	9/30/2006
Advanced Research Corporation	2001-3E-4601	Heat Assisted Magnetic Recording	Joint-Venture - Partner	Minnesota	11/1/2001	10/31/2006
Ansoft Corporation	2001-3E-4601	Heat Assisted Magnetic Recording	Joint-Venture - Partner	Pennsylvania	11/1/2001	10/31/2006
Cabot Corporation (formerly Superior MicroPowders, LLC)	2002-1 E-51 39	Nanoengineered Thermal Interfaces Enabling Next Generation Microelectronics	Joint-Venture - Partner	New Mexico	11/1/2002	10/31/2005
Cabot Corporation (formerly Superior MicroPowders, LLC)	1998-03-0002	Preparation and Fundamental Evaluation of Catalytic Materials for Energy Applications	Single Applicant	New Mexico	11/1/1 998	10/31/2001
Carbon Nanotechnologies, Inc.	2004-1C-6962	"Free Standing" Single-Wall Carbon-Nanotube Fuel Cell Electrode	Joint-Venture - Lead	Texas	10/1/2004	9/30/2007
Consultant	2000-1 B-4243	Advanced Wafer Inspection for Next-Generation Lithography	Subcontractor	Pennsylvania	11/1/2000	10/31/2004
Consultant	2001-3E-4601	Heat Assisted Magnetic Recording	Subcontractor	California	11/1/2001	10/31/2006

(Continued)

Organization	Project Number	Project Title	Participation Type	State	Start Date	End Date
Consultant	2001-3E-4601	Heat Assisted Magnetic Recording	Subcontractor	Minnesota	11/1/2001	10/31/2006
Corning Tropel Corporation (formerly Tropel Corporation)	2000-1B-4243	Advanced Wafer Inspection for Next-Generation Lithography	Joint-Venture - Partner	New York	11/1/2000	10/31/2004
Dendritech, Inc.	1998-06-0052	Ultra-Low Dielectric Constant Materials for Integrated Circuit Interconnects	Subcontractor	Michigan	4/1/1999	9/30/2002
Energy Beam Sciences, Inc.	2004-1E-6814	Development and Demonstration of a Multiple, High- Current-Density Shaped E-Beam Column With Independent Vector Beam Placement	Subcontractor	Massachusetts	10/1/2004	9/30/2006
Euxine Technologies	2001-3E-4601	Heat Assisted Magnetic Recording	Joint-Venture - Partner	Colorado	11/1/2001	10/31/2006
HIMARC Simulations	2002-3E-5853	Nano-Imprint Lithography Infrastructure for Low Cost Replication at the 65 nm Node and Beyond	Subcontractor	California	5/1/2004	4/30/2007
Intematix Corporation	2001-2C-4478	Rapid and Comprehensive Development of Advanced Dielectric Materials for Wireless Appplications	Single Applicant	California	10/1/2001	1/31/2004
InterScience, Inc.	2000-1B-4287	Digital Holographic Inspection of Semiconductor Devices	Joint-Venture - Partner	New York	11/1/2000	5/9/2006
Intevac, Inc.	2000-1B-4201	Deposition Source for Producing Super Lattice, Multilayer Thin Films to Enable Perpendicular Magnetic Recording	Single Applicant	California	11/1/2000	10/31/2003
JME, Inc.	1998-03-0027	Asymmetric Supercapacitor Based Upon Nanostructured Active Materials	Subcontractor	Ohio	3/1/1999	2/28/2002
Light Age, Inc.	2000-1B-4287	Digital Holographic Inspection of Semiconductor Devices	Joint-Venture - Partner	New Jersey	11/1/2000	5/9/2006

(Continued)

Organization	Project Number	Project Title	Participation Type	State	Start Date	End Date
Luxtera, Inc.	2002-1 E-4973	Nanophotonic Integrated Circuits for Telecommunications and Computing	Single Applicant	California	11/1/2002	10/31/2004
MEMS Optical, Inc.	2001-3E-4601	Heat Assisted Magnetic Recording	Joint-Venture - Partner	Alabama	11/1/2001	10/31/2006
Micro Magnetics, Inc.	2002-3E-5866	Spintronics-Based High-Resolution, Non-Invasive, and Ultrafast Metrology for the Semiconductor Industry	Single Applicant	Massachusetts	10/1/2003	9/30/2006
Micro Star Technologies	2004-1E-6249	High-Speed AFM-Based Platform for Quantitative Nanomechanical Measurements	Subcontractor	Texas	12/1/2004	9/30/2007
Microcell Corporation	2001-2C-4429	Fabrication of Fuel Cells from Microcell Fibers	Single Applicant	North Carolina	11/1/2001	10/31/2004
Molecular Imprints, Inc.	2002-3E-5853	Nano-Imprint Lithography Infrastructure for Low Cost Replication at the 65 nm Node and Beyond	Joint-Venture - Lead	Texas	5/1/2004	4/30/2007
Molecular Nanosystems, Inc.	2002-1 C-5223	Template Synthesis Platform for Nanostructu red Materials	Joint-Venture - Partner	California	11/1/2002	12/31/2005
Molecular Probe (formerly Quantum Dot Corporation)	2004-1 B-7002	Quantum Dots for Biomedical and Consumer Applications	Single Applicant	California	12/1/2004	6/30/2008
MRS, Inc.	2004-1E-6814	Development and Demonstration of a Multiple, High- Current-Density Shaped E-Beam Column With Independent Vector Beam Placement	Subcontractor	California	10/1/2004	9/30/2006
Multibeam Systems, Inc.	2004-1E-6814	Development and Demonstration of a Multiple, HighCurrent-Density Shaped E-Beam Column With Independent Vector Beam Placement	Single Applicant	California	10/1/2004	9/30/2006

(Continued)

Organization	Project Number	Project Title	Participation Type	State	Start Date	End Date
Nanocrystal Lighting Corp.	2004-1 E-701 1	Low Cost, High Efficiency Chip Scale LED Lamp	Joint-Venture - Partner	New York	11/1/2004	10/31/2007
Nanospectra Biosciences, Inc.	2004-1B-6963	The Seamless Detection and Treatment of Cancer With Near-Infrared Absorbing Nanoshells	Single Applicant	Texas	10/1/2004	6/30/2007
Neocera, Inc.	2001-5E-4850	Advanced Technology for Non-destructive, Localized, Dielectric Metrology of Future Generation Integrated Circuits	Single Applicant	Maryland	6/1/2002	5/31/2004
nLine Corporation	2000-1 B-4287	Digital Holographic Inspection of Semiconductor Devices	Joint-Venture - Lead	Texas	11/1/2000	5/9/2006
PixelVision, Inc.	2000-1B-4287	Digital Holographic Inspection of Semiconductor Devices	Joint-Venture - Partner	Oregon	11/1/2000	5/9/2006
Plug Power, Inc.	2002-3E-5836	Low Cost Fuel Cell System Technologies Development	Single Applicant	New York	12/1/2003	5/30/2006
Quantum Dot Corporation	2000-1A-4106	Blood "Fingerprinting": A First Step Toward Personalized Medicine	Joint-Venture - Partner	California	1/1/2001	6/30/2005
Standard MEMS, Inc.	2001-2E-4469	Assemblers for Nanotechnology Applications and Manufacturing: Enabling the Nanotechnology Era	Joint-Venture - Partner	Massachusetts	10/1/2001	12/16/2006
SurroMed, Inc.	2000-1A-4106	Blood "Fingerprinting": A First Step Toward Personalized Medicine	Joint-Venture - Lead	California	1/1/2001	6/30/2005
The Technological Group, Inc.	2004-1C-6962	"Free Standing" Single-Wall Carbon-Nanotube Fuel Cell Electrode	Subcontractor	Texas	10/1/2004	9/30/2007
Uncopiers, Inc.	2001-1E-4403	ACIM "Point*Suns": Concentrating Energy Through Silent Sound and Clean Water	Single Applicant	Kansas	9/1/2001	8/31/2004

(Continued)

Organization	Project Number	Project Title	Participation Type	State	Start Date	End Date
US Nanocorp, Inc.	1998-03-0027	Asymmetric Supercapacitor Based Upon Nanostructured Active Materials	Single Applicant	Connecticut	3/1/1999	2/28/2002
Zyvex Corporation	2001-2E-4469	Assemblers for Nanotechnology Applications and Manufacturing: Enabling the Nanotechnology Era	Joint-Venture - Lead	Texas	10/1/2001	12/16/2006
Medium-size Companies						
Cree, Inc	2004-1 E-701 1	Low Cost, High Efficiency Chip Scale LED Lamp	Joint-Venture - Lead	North Carolina	11/1/2004	10/31/2007
KLA-Tencor Corporation	2000-1B-4243	Advanced Wafer Inspection for Next-Generation Lithography	Joint-Venture - Lead	California	11/1/2000	10/31/2004
KLA-Tencor Corporation	2002-3E-5853	Nano-Imprint Lithography Infrastructure for Low Cost Replication at the 65 nm Node and Beyond	Joint-Venture - Partner	California	5/1/2004	4/30/2007
Photronics, Inc.	2002-3E-5853	Nano-Imprint Lithography Infrastructure for Low Cost Replication at the 65 nm Node and Beyond	Joint-Venture - Partner	Connecticut	5/1/2004	4/30/2007
Sarnoff Corporation	2000-1 B-4287	Digital Holographic Inspection of Semiconductor Devices	Joint-Venture - Partner	New Jersey	11/1/2000	5/9/2006
TIAX	2004-1E-7011	Low Cost, High Efficiency Chip Scale LED Lamp	Subcontractor	Massachusetts	11/1/2004	10/31/2007
Veeco Instruments, Inc.	2004-1 E-6249	High-Speed AFM-Based Platform for Quantitative Nanomechanical Measurements	Joint-Venture - Partner	California	12/1/2004	9/30/2007
Large-size Companies						
Celanese Ventures, GmbH	2004-1C-6962	"Free Standing" Single-Wall Carbon-Nanotube Fuel Cell Electrode	Subcontractor	New Jersey	10/1/2004	9/30/2007

(Continued)

Organization	Project Number	Project Title	Participation Type	State	Start Date	End Date
Corning IntelliSense Corporation	2002-1C-5223	Template Synthesis Platform for Nanostructu red Materials	Subcontractor	Massachusetts	11/1/2002	12/31/2005
Dow Chemical Company	1998-06-0052	Ultra-Low Dielectric Constant Materials for Integrated Circuit Interconnects	Joint-Venture - Lead	Michigan	4/1/1999	9/30/2002
Dow Chemical Company	2004-1E-6249	High-Speed AFM-Based Platform for Quantitative Nanomechanical Measurements	Joint-Venture - Lead	Michigan	12/1/2004	9/30/2007
Eveready Battery Company	1998-03-0027	Asymmetric Supercapacitor Based Upon Nanostructured Active Materials	Joint-Venture - Partner (Participant Type at Project Start)	Ohio	3/1/1999	2/28/2002
General Electric Company	2002-1C-5223	Template Synthesis Platform for Nanostructu red Materials	Joint-Venture - Lead	New York	11/1/2002	12/31/2005
General Electric Company	2002-1E-5139	Nanoengineered Thermal Interfaces Enabling Next Generation Microelectronics	Joint-Venture - Lead	New York	11/1/2002	10/31/2005
Honeywell International, Inc.	2001-2E-4469	Assemblers for Nanotechnology Applications and Manufacturing: Enabling the Nanotechnology Era	Joint-Venture - Partner	Washington	10/1/2001	12/16/2006
IBM, Almaden Research Center	1998-06-0052	Ultra-Low Dielectric Constant Materials for Integrated Circuit Interconnects	Joint-Venture - Lead	California	4/1/1999	9/30/2002
Johnson Matthey Fuel Cells, Inc.	2004-1C-6962	"Free Standing" Single-Wall Carbon-Nanotube Fuel Cell Electrode	Joint-Venture - Partner	Pennsylvania	10/1/2004	9/30/2007
Kellogg Brown & Root, Inc. (Halliburton)	2004-1C-6962	"Free Standing" Single-Wall Carbon-Nanotube Fuel Cell Electrode	Subcontractor	Texas	10/1/2004	9/30/2007

(Continued)

Organization	Project Number	Project Title	Participation Type	State	Start Date	End Date
Kelly Services	2004-1 C-6962	"Free Standing" Single-Wall Carbon-Nanotube Fuel Cell Electrode	Subcontractor	Texas	10/1/2004	9/30/2007
Motorola, Inc.	2004-1C-6962	"Free Standing" Single-Wall Carbon-Nanotube Fuel Cell Electrode	Joint-Venture - Partner	Arizona	10/1/2004	9/30/2007
Motorola, Inc. Motorola Labs	2002-3E-5853	Nano-Imprint Lithography Infrastructure for Low Cost Replication at the 65 nm Node and Beyond	Joint-Venture - Partner	Arizona	5/1/2004	4/30/2007
Motorola, Inc., Semiconductor Product Services	2002-1 E-4973	Nanophotonic Integrated Circuits for Telecommunications and Computing	Subcontractor	Texas	11/1/2002	10/31/2004
Seagate Technology, LLC	2001-3E-4601	Heat Assisted Magnetic Recording	Joint-Venture - Partner	California	11/1/2001	10/31/2006
Federal Laboratory						
Oak Ridge National Laboratory	2000-1 B-4287	Digital Holographic Inspection of Semiconductor Devices	Subcontractor	Tennessee	11/1/2000	5/9/2006
Non-Profit Organization						
Information Storage Industry Consortium (formerly National Storage Industry Consortium)	2001-3E-4601	Heat Assisted Magnetic Recording	Joint-Venture - Lead	California	11/1/2001	10/31/2006
Universities						
Carnegie Mellon University	2000-1B-4243	Advanced Wafer Inspection for Next-Generation Lithography	Joint-Venture - Partner	Pennsylvania	11/1/2000	10/31/2004
Carnegie Mellon University	2001-3E-4601	Heat Assisted Magnetic Recording	Joint-Venture - Partner	Pennsylvania	11/1/2001	10/31/2006

(Continued)

Organization	Project Number	Project Title	Participation Type	State	Start Date	End Date
Florida Atlantic University (Florida Atlantic Research Corporation)	1998-03-0027	Asymmetric Supercapacitor Based Upon Nanostructured Active Materials	Subcontractor	Florida	3/1/1999	2/28/2002
North Carolina State University	2001-2C-4429	Fabrication of Fuel Cells from Microcell Fibers	Subcontractor	North Carolina	11/1/2001	10/31/2004
Purdue University	2002-1E-5139	Nanoengineered Thermal Interfaces Enabling Next Generation Microelectronics	Subcontractor	Indiana	11/1/2002	10/31/2005
Rensselaer Polytechnic Institute	2001-2E-4469	Assemblers for Nanotechnology Applications and Manufacturing: Enabling the Nanotechnology Era	Subcontractor	New York	10/1/2001	12/16/2006
Stanford University	1998-06-0052	Ultra-Low Dielectric Constant Materials for Integrated Circuit Interconnects	Subcontractor	California	4/1/1999	9/30/2002
State University of New York (SUNY) at Binghamton	2002-1 E-51 39	Nanoengineered Thermal Interfaces Enabling Next Generation Microelectronics	Joint-Venture - Partner	New York	11/1/2002	10/31/2005
Texas A&M University	2004-1B-6963	The Seamless Detection and Treatment of Cancer With Near-Infrared Absorbing Nanoshells	Subcontractor	Texas	10/1/2004	6/30/2007
University of Arizona (Arizona Board of Regents)	2001-3E-4601	Heat Assisted Magnetic Recording	Joint-Venture - Partner	Arizona	11/1/2001	10/31/2006
University of Albany, Albany NanoTech	2002-3E-5836	Low Cost Fuel Cell System Technologies Development	Subcontractor	New York	12/1/2003	5/30/2006
University of California at Santa Barbara	2004-1 E-6249	High-Speed AFM-Based Platform for Quantitative Nanomechanical Measurements	Subcontractor	California	12/1/2004	9/30/2007

(Continued)

Organization	Project Number	Project Title	Participation Type	State	Start Date	End Date
University of Colorado	2002-1 E-5139	Nanoengineered Thermal Interfaces Enabling Next Generation Microelectronics	Subcontractor	Colorado	11/1/2002	10/31/2005
University of Maryland	1998-06-0052	Ultra-Low Dielectric Constant Materials for Integrated Circuit Interconnects	Subcontractor	Maryland	4/1/1999	9/30/2002
University of Maryland	2001-5E-4850	Advanced Technology for Non-destr-uctive, Localized, Dielectric Metrology of Future Generation Integrated Circuits	Subcontractor	Maryland	6/1/2002	5/31/2004
University of Maryland	2001-5E-4850	Advanced Technology for Non-destru-ctive, Localized, Dielectric Metrology of Future Generation Integrated Circuits	Subcontractor	Maryland	6/1/2002	5/31/2004
University of Maryland	2002-1C-5223	Template Synthesis Platform for Nanostructu red Materials	Subcontractor	Maryland	11/1/2002	12/31/2005
University of Minnesota	2001-3E-4601	Heat Assisted Magnetic Recording	Joint-Venture - Partner	Minnesota	11/1/2001	10/31/2006
University of Northern Texas	2001-2E-4469	Assemblers for Nanotechnology Applications and Manufacturing: Enabling the Nanotechnology Era	Subcontractor	Texas	10/1/2001	12/16/2006
University of Texas at Austin	2002-3E-5853	Nano-Imprint Lithography Infrastructure for Low Cost Replication at the 65 nm Node and Beyond	Joint-Venture - Partner	Texas	5/1/2004	4/30/2007
University of Texas at Dallas	2001-2E-4469	Assemblers for Nanotechnology Applications and Manufacturing: Enabling the Nanotechnology Era	Subcontractor	Texas	10/1/2001	12/16/2006
Washington State University	1998-06-0052	Ultra-Low Dielectric Constant Materials for Integrated Circuit Interconnects	Subcontractor	Washington	4/1/1999	9/30/2002

Appendix B. Advanced Technology Program Nanotechnology Awards for Nanoelectronic Applications by Innovation and Early-on Application Areas

Project Number	Lead Organization	Title	Award Type	Start Date	End Date	ATP Share, $ (est.)	Industry Share, $ (est.)
Nanotechnology Innovation Areas							
Devices and Systems							
2000-1B-4201	Intevac, Inc.	Deposition Source for Producing Super Lattice, Multilayer Thin Films to Enable Perpendicular Magnetic Recording	Single Applicant	11/1/2000	10/31/2003	$1,554,114	$1,554,114
2002-1E-4973	Luxtera, Inc.	Nanophotonic Integrated Circuits for Telecommunications and Computing	Single Applicant	11/1/2002	10/31/2004	$1,999,960	$2,193,728
2002-3E-5853	Molecular Imprints, Inc.	Nano-Imprint Lithography Infrastructure for Low Cost Replication at the 65 nm Node and Beyond	Joint-Venture	5/1/2004	4/30/2007	$17,623,118	$19,167,367
2004-1E-6814	Multibeam Systems Inc.,	Development and Demonstration of a Multiple, High- Current-Density Shaped E-Beam Column With Independent Vector Beam Placement	Single Applicant	10/1/2004	9/30/2006	$1,999,180	$708,300
2004-1B-6963	Nanospectra Biosciences, Inc.	The Seamless Detection and Treatment of Cancer With Near-Infrared Absorbing Nanoshells	Single Applicant	10/1/2004	6/30/2007	$2,000,000	$551,351
2000-1A-4106	SurroMed, Inc.	Blood "Fingerprinting": A First Step Toward Personalized Medicine	Joint-Venture	1/1/2001	6/30/2005	$5,595,485	$5,708,524

(Continued)

Project Number	Lead Organization	Title	Award Type	Start Date	End Date	ATP Share, $ (est.)	Industry Share, $ (est.)
Instrumentation and Metrology							
2004-1 E-6249	Dow Chemical Company	High-Speed AFM-Based Platform for Quantitative Nanomechanical Measurements	Joint-Venture	12/1/2004	9/30/2007	$6,560,714	$6,828,495
2000-1B-4243	KLA-Tencor Corporation	Advanced Wafer Inspection for Next-Generation Lithography	Joint-Venture	11/1/2000	10/31/2004	$6,716,870	$7,002,847
2002-3E-5866	Micro Magnetics, Inc.	Spintronics-Based High-Resolution, Non-Invasive, and Ultrafast Metrology for the Semiconductor Industry	Single Applicant	10/1/2003	9/30/2006	$2,000,000	$1,406,744
2001-5E-4850	Neocera, Inc.	Advanced Technology for Non-destructive, Localized, Dielectric Metrology of Future Generation Integrated Circuits	Single Applicant	6/1/2002	5/31/2004	$1,968,348	$1,410,453
2000-1B-4287	nLine Corporation	Digital Holographic Inspection of Semiconductor Devices	Joint-Venture	11/1/2000	5/9/2006	$9,395,604	$9,779,101
Manufacturing and Process Technologies							
2001-3E-4601	Information Storage Industry Consortium (formerly National Storage Industry Con)	Heat Assisted Magnetic Recording	Joint-Venture	11/1/2001	10/31/2006	$10,760,830	$10,838,696
2001-2C-4478	Intematix Corporation	Rapid and Comprehensive Development of Advanced Dielectric Materials for Wireless Appplications	Single Applicant	10/1/2001	1/31/2004	$2,000,000	$1,169,688
2004-1 B-7002	Quantum Dot Corporation	Quantum Dots for Biomedical and Consumer Applications	Single Applicant	12/1/2004	6/30/2008	$2,000,000	$888,524

(Continued)

Project Number	Lead Organization	Title	Award Type	Start Date	End Date	ATP Share, $ (est.)	Industry Share, $ (est.)
2001-1E-4403	Uncopiers, Inc.	ACIM "Point*Suns": Concentrating Energy Through Silent Sound and Clean Water	Single Applicant	9/1/2001	8/31/2004	$2,000,000	$310,730
2001-2E-4469	Zyvex Corporation	Assemblers for Nanotechnology Applications and Manufacturing: Enabling the Nanotechnology Era	Joint-Venture	10/1/2001	12/16/2006	$12,170,000	$12,245,945
Nanomaterials							
1998-03-0002	Cabot Superior MicroPowders (formerly Superior MicroPowders)	Preparation and Fundamental Evaluation of Catalytic Materials for Energy Applications	Single Applicant	11/1/1998	10/31/2001	$2,000,000	$1,842,811
2004-1C-6962	Carbon Nanotechnologies, Inc.	"Free Standing" Single-Wall Carbon-Nanotube Fuel Cell Electrode	Joint-Venture	10/1/2004	9/30/2007	$3,616,054	$3,763,653
2004-1E-7011	Cree, Inc	Low Cost, High Efficiency Chip Scale LED Lamp	Joint-Venture	11/1/2004	10/31/2007	$3,383,865	$3,567,330
1998-06-0052	Dow Chemical Company	Ultra-Low Dielectric Constant Materials for Integrated Circuit Interconnects	Joint-Venture	4/1/1999	9/30/2002	$8,556,629	$9,049,142
2002-1C-5223	General Electric Company	Template Synthesis Platform for Nanostructu red Materials	Joint-Venture	11/1/2002	12/31/2005	$2,833,996	$2,949,672
2002-1E-5139	General Electric Company	Nanoengineered Thermal Interfaces Enabling Next Generation Microelectronics	Joint-Venture	11/1/2002	10/31/2005	$3,506, 139	$3,761,381
200 1-2C-4429	Microcell Corporation	Fabrication of Fuel Cells from Microcell Fibers	Single Applicant	11/1/2001	10/31/2004	$1,997,923	$404,033
2002-3E-5836	Plug Power Inc.	Low Cost Fuel Cell System Technologies Development	Single Applicant	12/1/2003	5/30/2006	$1,879,586	$1,973,957

(Continued)

Project Number	Lead Organization	Title	Award Type	Start Date	End Date	ATP Share, $ (est.)	Industry Share, $ (est.)
1998-03-0027	US Nanocorp, Inc.	Asymmetric Supercapacitor Based Upon Nanostructured Active Materials	Single Applicant	3/1/1999	2/28/2002	$441,000	$459,000
Early-on Application Areas							
Alternative Energy							
1998-03-0002	Cabot Superior MicroPowders (formerly Superior MicroPowders)	Preparation and Fundamental Evaluation of Catalytic Materials for Energy Applications	Single Applicant	11/1/1998	10/31/2001	$2,000,000	$1,842,811
2004-1C-6962	Carbon Nanotechnologies, Inc.	"Free Standing" Single-Wall Carbon-Nanotube Fuel Cell Electrode	Joint-Venture	10/1/2004	9/30/2007	$3,616,054	$3,763,653
200 1-2C-4429	Microcell Corporation	Fabrication of Fuel Cells from Microcell Fibers	Single Applicant	11/1/2001	10/31/2004	$1,997,923	$404,033
2002-3E-5836	Plug Power Inc.	Low Cost Fuel Cell System Technologies Development	Single Applicant	12/1/2003	5/30/2006	$1,879,586	$1,973,957
1998-03-0027	US Nanocorp, Inc.	Asymmetric Supercapacitor Based Upon Nanostructured Active Materials	Single Applicant	3/1/1 999	2/28/2002	$441,000	$459,000
Bioelectronics							
2004-1 B-6963	Nanospectra Biosciences, Inc.	The Seamless Detection and Treatment of Cancer With Near-Infrared Absorbing Nanoshells	Single Applicant	10/1/2004	6/30/2007	$2,000,000	$551,351
2004-1 B-7002	Quantum Dot Corporation	Quantum Dots for Biomedical and Consumer Applications	Single Applicant	12/1/2004	6/30/2008	$2,000,000	$888,524
2000-1A-4106	SurroMed, Inc.	Blood "Fingerprinting": A First Step Toward Personalized Medicine	Joint-Venture	1/1/2001	6/30/2005	$5,595,485	$5,708,524

(Continued)

Project Number	Lead Organization	Title	Award Type	Start Date	End Date	ATP Share, $ (est.)	Industry Share, $ (est.)
Materials Discovery							
2004-1 E-6249	Dow Chemical Company	High-Speed AFM-Based Platform for Quantitative Nanomechanical Measurements	Joint-Venture	12/1/2004	9/30/2007	$6,560,714	$6,828,495
2001-2C-4478	Intematix Corporation	Rapid and Comprehensive Development of Advanced Dielectric Materials for Wireless Appplications	Single Applicant	10/1/2001	1/31/2004	$2,000,000	$1,169,688
Mixed-function Electronic Systems							
2004-1E-7011	Cree, Inc	Low Cost, High Efficiency Chip Scale LED Lamp	Joint-Venture	11/1/2004	10/31/2007	$3,383,865	$3,567,330
2002-1 C-5223	General Electric Company	Template Synthesis Platform for Nanostructu red Materials	Joint-Venture	11/1/2002	12/31/2005	$2,833,996	$2,949,672
2002-1E-5139	General Electric Company	Nanoengineered Thermal Interfaces Enabling Next Generation Microelectronics	Joint-Venture	11/1/2002	10/31/2005	$3,506,139	$3,761,381
2001-3E-4601	Information Storage Indu-stry Consorium (formerly National Storage Industry Con)	Heat Assisted Magnetic Recording	Joint-Venture	11/1/2001	10/31/2006	$10,760,830	$10,838,696
2000-1B-4201	Intevac, Inc.	Deposition Source for Producing Super Lattice, Multilayer Thin Films to Enable Perpendicular Magnetic Recording	Single Applicant	11/1/2000	10/31/2003	$1,554,114	$1,554,114

(Continued)

Project Number	Lead Organization	Title	Award Type	Start Date	End Date	ATP Share, $ (est.)	Industry Share, $ (est.)
2002-1 E-4973	Luxtera, Inc.	Nanophotonic Integrated Circuits for Telecommunications and Computing	Single Applicant	11/1/2002	10/31/2004	$1,999,960	$2,193,728
2001-2E-4469	Zyvex Corporation	Assemblers for Nanotechnology Applications and Manufacturing: Enabling the Nanotechnology Era	Joint-Venture	10/1/2001	12/16/2006	$12,170,000	$12,245,945
Semiconductor Nanomanufacturing							
1998-06-0052	Dow Chemical Company	Ultra-Low Dielectric Constant Materials for Integrated Circuit Interconnects	Joint-Venture	4/1/1999	9/30/2002	$8,556,629	$9,049,142
2000-1B-4243	KLA-Tencor Corporation	Advanced Wafer Inspection for Next-Generation Lithography	Joint-Venture	11/1/2000	10/31/2004	$6,716,870	$7,002,847
2002-3E-5866	Micro Magnetics, Inc.	Spintronics-Based High-Resolution, Non-Invasive, and Ultrafast Metrology for the Semiconductor Industry	Single Applicant	10/1/2003	9/30/2006	$2,000,000	$1,406,744
2002-3E-5853	Molecular Imprints, Inc.	Nano-Imprint Lithography Infrastructure for Low Cost Replication at the 65 nm Node and Beyond	Joint-Venture	5/1/2004	4/30/2007	$17,623,118	$19,167,367
2004-1E-6814	Multibeam Systems Inc.,	Development and Demonstration of a Multiple, High- Current-Density Shaped E-Beam Column With Independent Vector Beam Placement	Single Applicant	10/1/2004	9/30/2006	$1,999,180	$708,300
2001-5E-4850	Neocera, Inc.	Advanced Technology for Non-destructive, Localized, Dielectric	Single Applicant	6/1/2002	5/31/2004	$1,968,348	$1,410,453

(Continued)

Project Number	Lead Organization	Title	Award Type	Start Date	End Date	ATP Share, $ (est.)	Industry Share, $ (est.)
		Metrology of Future Generation Integrated Circuits					
2000-1B-4287	nLine Corporation	Digital Holographic Inspection of Semiconductor Devices	Joint-Venture	11/1/2000	5/9/2006	$9,395,604	$9,779,101
200 1-1 E-4403	Uncopiers, Inc.	ACIM "Point*Suns": Concentrating Energy Through Silent Sound and Clean Water	Single Applicant	9/1/2001	8/31/2004	$2,000,000	$310,730

End Notes

[1] Sizing Nanotechnology's Value Chain, Lux Research Report, October 2004.
[2] http://physics.nist.gov/Divisions/Div841/Gp3/cnst_home.html.
[3] *Accomplishments in Nanotechnology,* NIST Special Publication: 1052, August 2006.
[4] ATP defines a "company" as a for-profit business that is independently owned and operated and is organized for profit. A "small company" is a business with fewer than 500 employees. A "large company" is any business, including any parent company plus related subsidiaries, having annual revenues that places it within the Fortune 500 listing of companies for the year the award is issued. A "medium company" is a business that falls in between a small and large company.
[5] Participants represent all organizations, including subcontractors, and may include multiple instances of involvement by the same organization.
[6] A nanometer is one-billionth of a meter (10xE-9 m). Novel and differentiating properties and functions of nanomaterials and nanoscale devices arise at this critical length scale of matter (http://www.nano.gov/html/facts/whatIsNano.html).
[7] http://www.sric-bi.com/Explorer/NGT-BE.shtml.
[8] The National Nanotechnology Initiative – Supplement to the President's 2006 Budget, National Science, Engineering, and Technology Subcommittee, Committee on Technology, National Science and Technology Council, March 2005.
[9] http://jazz.nist.gov/atpcf/prjbriefs/prjbrief.cfm?ProjectNumber=00-00-4973.
[10] "Ride the Light," by Elizabeth Corcoran, Forbes.com, 4/11/2005, http://www.forbes.com/business/free_forbes/2005/0411/068.html.
[11] http://www.mel.nist.gov/proj/nm.htm.
[12] http://jazz.nist.gov/atpcf/prjbriefs/prjbrief.cfm?ProjectNumber=00-00-5866.
[13] http://jazz.nist.gov/atpcf/prjbriefs/prjbrief.cfm?ProjectNumber=00-00-4403.
[14] http://jazz.nist.gov/atpcf/prjbriefs/prjbrief.cfm?ProjectNumber=00-00-7011.

CHAPTER SOURCES

The following chapters have been previously published:

Chapter 1 – This is an edited, reformatted and augmented version of National Science Foundation Workshop Report on Re-Engineering Basic Research, dated March 16-19, 2008.

Chapter 2 – This is an edited, reformatted and augmented version of National Institute of Standars and Technology, dated May 20-22, 2008.

Chapter 3 – This is an edited, reformatted and augmented version of National Institute of Standars and Technology, dated October 2006.

INDEX

G

H

I

Q

R

S